Whose Brave New World?

The Information Highway and the New Economy

WHOSE BRAVE NEW

HEATHER MENZIES

world?

the information highway and the new economy

BETWEEN THE LINES
Toronto, Canada

Published by
Between The Lines
720 Bathurst Street, #404
Toronto, Ontario M5S 2R4
Canada

Fifth printing, November 2001

Cover and text design by Gordon Robertson
Cover Photomontage by Richard Slye
Author photo by Jayne Balharrie
Printed in Canada by union labour

Between The Lines gratefully acknowledges financial assistance from the Canada Council, the Ontario Arts Council, and the Canadian Heritage Ministry.

Canadian Cataloguing in Publication Data

Menzies, Heather, 1949-
Whose brave new world? : the information highway and the new economy

Includes index.
ISBN 1-896357-02-4

1. Technological unemployment. 2. Information society.
3. Economic history - 20th century. I. Title

HD6331.M45 1996 331.137'042 C96-930135-9

For my Mum, Anne Menzies

Contents

Acknowledgements xi

Introduction xiii

Part I The Issues: Deficits in the Context of Globalization 1

1. **Behind the Silicon Curtain: Perception Management and the Adjustment Agenda** 3

The Information Highway as Restructuring Agent 7
The Digital Divide and the New Inequalities 10
Getting Grounded in What's Happening to People 14

2. **The Chip and Programmed People: The Real World of Restructuring** 20

The Disappearance of the Restructuring Debate 23
Opening the Black Box of Technology 26
Chicken Barns and Push Buttons: The Story of Restructuring So Far 29
Free Trade, Deficit-Cutting, and Other Adjustments 37
Time, Moore's Law, and the Technological Dynamo 42
The Technological Discourse and Murphy's Law 45

Part II Background: Social and Cultural Transformations 49

3. **Hype and the Highway: Virtual Corporations and the Agile Workforce** 51

The Information Highway, IHAC,
and Technological Restructuring 52
Restructuring and Social Impacts 59
in offices 60
in insurance 63
in hospitals 65
in retail services 69
Virtual Corporations and the Virtual New Economy 72

4. **Across the Digital Divide: Manufacturing as Global Agility** 80

The Stepping Stones to Quick-Response Agility 83
Computer-integrated manufacturing 83
Just-in-time (*JIT*) manufacturing 84
Lean production 84
Agile or "flexible" manufacturing 85
Case-Study Evidence 89
Resources 89
The garment industry 90
Railways 92
Telecommunications 93
Automobiles 97
Aerospace 99
TQM and Integrating People into
the Corporate Vision 102
TQM and Tacit Knowledge 105

5. **Panopticons and Telework: The New Cybernetics of Labour** 109

Digitizing Memory 111
Job Title: Traffic Operator Position 113
Call-Centres and Telework 114
Computer-Monitoring
and Performance Measurement 117
Training for Compliance 122
Taylorism and the Panopticon 125
Repetitive Strain Injury
and Layoff Survivor Syndrome 129

Part III Response: Restructuring for People 131

6. A Plea for Time—A Plea for Our Times 133

The Case of the Midland Operators 134
Towards a New Critical Discourse 137

7. On Our Own Terms 143

The Bias of Communication:
Highway Transmission or
Communing and Community 145
A Social Contract for the Information Highway 150
Virtual Unions and Coalition-Building 155
Coalition on working hours 156
Labour standards 158
Machine census and "head tax" 158
The "reverse strike" and workfare 161
A Constitution for the
Information Highway and the New Economy 162

Notes 165

Index 181

Acknowledgements

THE WRITING OF A BOOK is a webwork emerging out of other webworks, so it's impossible to quantify all the elements which made it possible, nor to thank all the people whose assistance, support, and inspiration made it happen. Still, I will try. Some names spring easily to mind: my editor, Robert G. Clarke, whose steadiness, intelligence, and careful attention to detail helped prepare my sometimes windy words for publication. My thanks also to Jamie Swift, who read the manuscript in draft form and commented most helpfully on it; and to Carol Lane who did some essential research for me. Also, thanks to Bruce Allen, Robert Babe, Ellen Balka, Marc Bélanger, Bruce Campbell, Janet Dassinger, Karen Hadley, Roy Hanna, Theresa Johnson, Nancy Jollicourt, Robert Mullaly, Margaret Oldfield, Bruce Roberts, Andrew Reddick, Andrew Sharpe, Laura Sky, Gail Stewart, Joan Weinman, and Lynda. I also want to thank two people whose writings helped crystallize my thinking. Dorothy Smith for her book, *The Conceptual Practices of Power*; and John Ralston Saul for his *Voltaire's Bastards* and the 1995 CBC Massey Lectures. I also wish to acknowledge the generous support of the Ontario Arts Council and the Canada Council.

A special thanks to Ursula Franklin, whose writings (including *The Real World of Technology*), wisdom, and personal support inform my writing. And, finally, ongoing thanks to my partner Miles Burton and my son Donald for all their loving support.

Introduction

I JOINED the debate on technological restructuring some fifteen years ago, when computerization was just beginning its march across the social and economic landscape. I wanted to find out what the changes meant for people—this massive transformation over which most had little or no control. In Ste. Agathe, Quebec, I met a group of women who had operated the telephone exchange there all their adult lives. Now they were unemployed because Bell Canada had computerized the switching of both local and long-distance calls in a way that made them abruptly "redundant."

"They treated the machines better than they treated us," the women told me. The exchange equipment was redeployed, while most of the women were sent home. They were sent away with roses—plastic roses, "fifteen-cent-a-piece plastic roses," they told me bitterly.

I wrote down their stories, and others I found, and brought them back to the research institute that would publish my work. But the debate by then ran mostly on numbers, on large-scale statistical trends. To be recognized as having something to say, I had to come up with projections. So I did. The women I had met turned into X number of clerical workers likely to be displaced in Y years unless innovative new work opened up to replace the Z jobs being lost to computerization. In this account the sense of human choice in the design and use of technology—to enhance people's work or to replace them—had disappeared. The women's humanity disappeared too as I parked their stories at the door and objectified them into categories. I didn't sense my hand in their disappearance.

I also didn't imagine that I might have betrayed them. Yet in adopting the official language of the discourse, I translated them from historical agents with a moral claim arising from their lived reality into objects of the discourse's reality, a context in which they had no claim at all. They became data in the never-never land of futures projections; their reality was gone.

In a sense, I betrayed myself too. I wasn't speaking for myself anymore. I was speaking for the discourse. I was speaking for what its terms of reference designated as real: competitive productivity increases, new skills, and a few inevitable redundancies—not people with a stake in continuing their involvement in a community's economy and in something they knew, understood, and did well.

I exchanged my own agency as a writer—what Marshall McLuhan has called the artist as living "probe"—for that of scientific "expert" in the discourse. But I was an expert only within its system of delegated authority, including its system for designating what was relevant and real. People serving communities weren't real. Skill-sets in a competitive corporate economy were real.

It was heady and exciting for a while, and a flurry of media attention assured me that what I had to say was important. Yet I was fundamentally at odds with myself. There I was, urging that the restructuring process be informed by people, human values, and social justice, not by the systems values of the corporate economy that had dominated so far. Yet I had adopted the systems language and its objectifying grammar to make my point.

At first, too, I was persuaded to view restructuring simply as an economic issue. So it was enough simply to point out what was necessary in training women (in *Women and the Chip*) and men (in *Computers on the Job*) if they were to adjust to the new technologies. Seven years later (in *Fastforward and Out of Control*), I realized the issue was a social and political matter. Now, as it intersects with the unfolding information highway, I can see that it is even bigger than that. The issue of technological restructuring is altering not just our jobs and our work, but our language, consciousness, and identity.

Now, too, I realize the importance of language in the construction of public perception. If people are reduced to objects in the discourse about restructuring, and if the global economy is portrayed as primary subject and agent, then people can be more readily treated as objects in reality. If their humanity is silenced in the discourse informing public perception, if the social realities and the social values of restructuring are marginalized, it is harder to come to terms with restructuring as the profoundly social, political, and even moral issue it is.

Given the trends, we could be headed for a cybernetic brave new world, a digital retread of feudalism, with electronic moats between the new haves and have-nots, between those wielding vast computer power and control, and those on the receiving, programmed end of the changes.

But we will stay on this road only if present trends are allowed to continue. We can turn things around. We can gain control of the restructuring process, if we can renew a purposeful public debate about it as something to be negotiated, as a struggle between very different values: the logic of competition versus the logic of community; the logic of machines and machine efficiency versus the logic of people trying to make a life for themselves and participate meaningfully in their society. To do that, though, we will need a new critical discourse on technology, grounded in the social context and in the language of human experience.

I will spend a lot of time through this book focusing on managed public perception and the various ways in which this practice operates, through the mass media, advertising, and the official discourse on restructuring as well as through the ad motifs in corporate cultural training and "total quality management" programs. Equally, though, I will also speak for myself.

I have gathered many facts and statistics on what's happening, but I've also dwelled on the trends as human experience. In doing this I'm hoping not only to provide useful information, but also to foster a feeling for what's happening. I hope this helps all of us find a way to act on whatever level and in whatever way we can to challenge the current direction and to turn it around.

The postindustrial society can and should be inclusive and participatory, just and democratic—truly an information and knowledge society. And it might turn out that way if we can break the anaesthetizing spell of the official adjustment agenda and the discourse around it. It all depends on our willingness (and daring) to find our voices as people (instead of as experts or bystanders or inert victims) and to put people back at the centre of the discussion.

I've tried to approach this book on both a global and a local level. At the local level, I hope it helps people to gain control over the technologies in their workplaces so they can use them to extend the scope of what they do. At a more global level, I hope it will open up a broader, more inclusive, and participatory vision around the information highway and the new postindustrial economy, and help people identify what's needed to achieve it.

Part 1 sets out the main argument—namely that social-spending cuts and deficit reduction are not the key issues of the day; they are only part of the larger agenda associated with technological restructuring. Those issues, plus the late 1980s debates about free trade and deregulation, represent the administrative and social-policy follow-up to the technological restructuring going on in local workplaces and throughout the economy. That restructuring, I believe, is geared to the interests and needs not of people, but of corporate economic renewal and global expansion.

The first chapter lays out the main themes and describes the disintegration of society being caused by technological restructuring. The shift to the information highway is the final phase of a twenty-year-long restructuring of the corporate economy to create a global systems economy. The stages in this massive restructuring run from the automation and integration phases of the 1980s to the networking and information highway phases of the 1990s. This restructuring has involved a massive erosion, deskilling, and demeaning of work and set the stage for a centralization of control through computers and a new regime of computerized control in work and society at large.

Part II provides background and critical evidence about, for instance, the link between restructuring and a high unemployment rate, jobless economic growth, and a two-tiered labour force with a new near-underclass of people who work only as operatives of computer systems that control and monitor everything they do. Part III suggests what we can do about it all.

One final note. This book began as an attempt to revise and update *Fastforward and Out of Control*, which went out of print six months after it was released in 1989, despite being on *The Globe and Mail*'s bestseller list. But so much has changed since then that it has turned into a whole new book. Still, some material from the earlier book has been recycled, though in an updated form.

PART 1

THE ISSUES

Deficits in the Context of Globalization

1

Behind the Silicon Curtain: Perception Management and the Adjustment Agenda

"Press this button and we'll give you $10." That's what a job's going to be.
– Donald Burton (age thirteen), Ottawa

They can talk about all their paradigms. But there's nothing new here. We've got people who've got to eat, who've got to put a roof over their heads. That's the day-to-day reality here.
– Greg MacKenzie, Glace Bay, Nova Scotia

The new media are not bridges between man and nature. They are nature.
– Marshall McLuhan

ON A BITTER JANUARY DAY in 1995, an estimated twenty-six thousand people lined up to apply for a rumoured third shift of jobs at a General Motors auto plant in Oshawa, Ontario. The *Toronto Star* ran a front-page headline, plus photos and quotes from people who had travelled from as far away as Regina. Some had camped out overnight and froze their toes in a desperate bid for a well-paying, full-time job. The next day the paper followed up with more front-page coverage and stories inside.

Pictures and stories ran in other newspapers across the country and on the television news. By contrast *The Globe and Mail*—Canada's "national newspaper," an agenda-setter for Question Period in Parliament, and a pillar of public discourse—ran only a brief item inside its business section on the first day, and nothing the second day. Instead it featured a front-page story about United Parcel Service transferring between 800 and 850 jobs to call-centres in New Brunswick. On January 12, the third day after the GM event, the *Globe* ran another front-page story on the UPS move, another story inside, and an editorial headlined "Utopia in the Boardroom." The *Globe* kept that story alive for an astounding nine days, including columns, op-ed-page pieces, cartoons, and other stories related to the subject of call-centres.

The message in *The Globe and Mail* was jobs, jobs, jobs, not unemployment and underemployment. The effect for its readers was the disappearance of the most dramatic sign of the times since the image of men riding the rails had captured the desperate despair of the Depression of the 1930s. What has sometimes been called the Fordist social contract, which saw decent incomes distributed widely to a population more or less fully employed in mass production and mass consumption, has collapsed. Some 20 per cent of the adult population is either unemployed or underemployed. Yet often it feels as though none of this is happening, none of it is real.

It reminds me of Orwell's *1984*: in particular, when Winston Smith hears a woman singing in the yard outside the window of the hideaway where he and Julia are snared by Big Brother; yet he can't open the window and save himself by joining this woman. He can't ally himself with her singing, sweating humanity.

When I read Orwell's novel I couldn't fathom how Winston could be that cut off from the Proles when they were so near to him all along. He was that closed off inside the world of Newspeak projected from the large telescreen on his apartment wall, with its double-plus good economic indicators and its constant exhortations to work harder, to make the necessary sacrifices. Plus the Proles were so different that he couldn't identify with them. They weren't on a dental plan, so they had bad teeth. They had no jobs—at least not in the core of the formal economy—so they didn't shave and get haircuts. They lacked critical analysis, at least as he was expecting it. They were without ambition, he concluded. They were pathetically content to languish away on welfare.

Now I understand how two realities can co-exist yet exclude each other. Because it's happening here in the 1990s, and it seems equally immobilizing. On the one hand there's the official reality projected by the major institutions of public information. This is a disembodied reality of government spending cuts and deficit-cutting targets plus a droning litany of "adjustments" to globalization and "the new economy." It is a reality named by experts and made both meaningful and normal by larger-than-life anchorpersons, columnists, and newspaper editorials. And it is reinforced daily on the telescreens of the mass media, programming us by the millions of our plugged-in minds.

Meanwhile, outside the window, society is staggering and falling apart.

While the Royal Bank joined Bell Canada in what the media playfully called "the Billion Dollar Club" with 1994 profits of over $1 billion, while its chief executive officer took the "salary crown" of the major Canadian banks with an income of $2.65 million, and while the Ontario Conservatives made a $900,000 profit from a single fund-raising dinner in April 1995, Toronto's Daily Bread Food Bank was feeding an average of 150,000 people, including 60,000 children per month, through 1994. The food bank's "clients," as they're called these days, included professionals (14 per cent) as well as office workers (19 per cent) with lots of computer skills.

The rich are getting richer and more powerful. The poor and the middle class are being flattened. Some 40 per cent of Canadians describe themselves as "economically distressed."[1] Record numbers of people are going bankrupt or losing homes in mortgage foreclosures. The official unemployment rate is just under 10 per cent, although the real rate, including those who have given up and those having to settle for underemployment in part-time, temporary jobs, is considered to be well over 20 per cent.

In 1993, the first year of "recovery" after the recession of the early 1990s, Canada's poor increased by five hundred thousand people. Worse, poverty had become increasingly concentrated among young people either working or looking for work.

"Not since the Depression of the 1930s have so many young families been at such risk of economic insecurity," said a report on youth poverty prepared by three Toronto social-service agencies.[2] In the mid-1990s nearly half of all young people (under twenty-five) are working part-time, compared to only a quarter of them in 1980. They not only earn less than half what full-time people earn; most have no benefits such as unemployment insurance, maternity leave, or a dental plan, let alone job security. In 1995, two years into the so-called recovery, the unemployment rate was still just a fraction below 10 per cent (9.8 per cent). The unemployment rate among fifteen- to twenty-four-year-olds was 17 per cent.

By 1993 only half of Canadian breadwinners had a full-time job, although a rising proportion of those who did were working fifty-hour weeks or more.[3] As for the great GM job lottery in Oshawa, twelve months later not one of the people who had applied for jobs had got one. Jobs—especially full-time jobs—are disappearing in every economic sector from resources through manufacturing to services, with public services decimated as much as commercial and retail services. The federal government announced that it would be laying off forty-five thousand public servants. Bell Canada said it would lay off ten thousand, and the banks are expected to drop thirty-five thousand employees within the next ten years.[4]

There is some job growth in challenging, high-technology areas—such as in networking systems and software, computer graphics, printing, and computer animation—and, tellingly, also in security: security-guard employment doubled between 1971 and 1991.[5] But the world of high tech is not a jobs cornucopia. In Ottawa, the high-tech capital of Canada, the sector accounts for only thirty-six thousand jobs, in part because its own inventions, such as automating and expert-level software, tend to cannibalize the new work that high technology does create.

When jobs do emerge, most of them are part-time, term, temporary, or contract—jobs that often won't support a decent single life, let alone a family. In New Brunswick, where building the information highway infrastructure has been the McKenna government's industrial strategy, most of the "high-technology" jobs have been in call-centres: in silicon work cells, where people are employed part-time and at close to minimum wage to supply rote responses to 1-800 calls. Their work is defined and monitored by computer systems they can't see or talk back to.

By 1994 the average family required two incomes to match its 1980 earnings. More couples are now moonlighting—with a fifth working ninety or more hours a week.[6] Meanwhile, nearly a third of Canadian workers worry that they don't have enough time with family or friends. A quarter feel stressed by work, especially by the increased pace encouraged if not demanded by the use of computers, faxes, and other technology. Repetitive strain injury has become the "occupational disease of the '90s," joined by a new condition called "layoff survivor syndrome," in

which people combine guilt at still holding a job wit
whether they too will be axed.[7]

According to an Ontario hospital study, "Nurses are
ing their job now that there's so much tension on the f
you even have shifts and people on the same shifts fighting each other."[8]

Deficit-cutting and social spending cuts are only part of the reason for all this misery. Technological restructuring is the larger reason—the real story that needs to be discussed, that calls for a response. In a process that goes back decades, national machine-based industrial economies are being restructured into a globally networked, computerized postindustrial economy. Privatizing public services, cutting social programs, levelling social standards, and impoverishing the working population are merely attendant adjustments.

The Information Highway as Restructuring Agent

The information highway is emerging as the axis of this new economy. The highway is a webwork of powerful (high-capacity) computer-communications networks capable of handling everything from video to voice, text to computer data and graphics, interchangeably, interactively, and at lightning speeds. Including private corporate networks, the Internet, and conventional cable broadcast and communications carriers now venturing into multimedia, the information highway is becoming the site of business deals, of money changing hands. It is the place where work is dispatched to new global and local labour markets, where work itself is done and supervised, and where value is said to be added. The restructuring associated with it is transforming the foundations of society and all its major institutions, by digitizing them and, increasingly, transforming them into extensions of remote information systems and service suppliers. These service and infrastructural suppliers ply the new information highway in much the same way as eighteenth-century merchant traders once roamed the old high seas.

The restructuring represents a staggering social transformation, equivalent to that of the Industrial Revolution. Then, in the nineteenth century, the context of the economy shifted from homes and household workshops to factories and offices. Now it is shifting to the infrastructures of the information highway. Work in every sector of the economy, from mining to manufacturing to the giving and receiving of services, is being transformed into extensions and expressions of management information systems. Not only are work and economic activity being digitized—that is, redefined in terms that computers can program and remotely control—but also, in the current networking phase of this transformation, economic activity as a whole is being scaled up to a standard that only the largest and most capital-intensive corporations and corporate economies can match.

In the mad scramble to keep up, to scale up, to step up the pace and to get there first, people are being left behind in every sense of the word, from the larger level of governance to the local level of employment and personal involvement. As a Conference Board of Canada report mildly put it, "Canadian businesses have used capital investment to satisfy their need to increase productivity at the expense of employment growth."[9]

The "downsizing" of organizations is a one-sided affair. People are being downsized, but the use of technology is growing in size and complexity. Check it out: count the burgeoning plethora of computers, data lines, and software packages crowding budget priority lists while people are declared "surplus." Take the federal government, for example. While Ottawa announced plans to lay off forty-five thousand public servants, it continued to spend billions of dollars a year on information technology: $3.1 billion in 1992-93; $3.5 billion in 1993-94; and $3.6 billion in 1994-95.[10] That sum could have employed ninety thousand public servants in 1995, at the average public service salary of $40,000 per person. Of course, the government can't simply get rid of all its computers, communications lines, and consultants; but consider the implications if it had shared the downsizing burden. If it had reduced its information-technology expenditures by a modest 10 per cent over two years from its 1992-93 levels, the money it would have saved could have allowed over twenty thousand people to keep on providing public services—close to half the total being cut in the Draconian downsizing announced in 1995.

In August 1995 the government announced plans to replace an additional 3,500 jobs in 150 employment centres across the country with 400 computerized kiosks. And so it goes. The National Trustco company acknowledged plans to cut two hundred employees after installing a $10 million computer system that will allow it to trim labour costs and boost its profit margin.[11] In every sector of the economy, machines are replacing people. Machine intelligence is replacing human intelligence, judgement, and discretion. Human infrastructures managed by local knowledge and personal involvement are being replaced by technological infrastructures managed through remote-control operating systems.[12] And this is happening at a seemingly exponential rate as the costs of technology plummet and the economies of scale expand. Corporate and government managers, controlling the purchase and use of the new technologies, are choosing to use them not to augment and expand what people do, but to replace them and to diminish and even control what they do.

This massive restructuring is closely linked to some dramatic new developments in the so-called labour market: protracted high levels of unemployment even in times of economic growth and record-breaking profits; rising levels of underemployment; and a polarization of the workforce into the working rich and the working poor. Technological restructuring—notably in how it is being managed—is possibly the most important single factor behind these changes. It is turning good full-time jobs into lousy part-time, shift, and temporary jobs, at what seems like an escalating rate. It is also hollowing out the middle ranks of administrative, sales, and service jobs, as well as professional, management, and skilled-trades work. Through computerization, work that had involved human judgement, intuition, skill, and autonomous decision-making is being transformed into an expression of computer logic: software does the important thinking and deciding, and it controls what can and cannot be done, or said, or given and received by way of service to the public.

The term "McJob" aptly applies to this new breed of computer-defined work, because the McDonald's hamburger chain was a pioneer in this field. The essence of this abbreviated form of work is that the computer (or the management information systems inside the computer) does all the thinking, organizing, and planning. Furthermore, the tasks are so completely determined and controlled by the computer system that job performance can be measured and monitored by the system itself.

The computer's simplification and control of work have made possible the replacement of full-time staff with part-time, temporary McJobs in every economic sector, from goods production to the provision of public, commercial, and personal services. More and more people are being marginalized in these computer-defined, computer-controlled jobs, if they are not being excluded altogether. Bureaucrats call this growing new category of peripheralized, marginalized McJobbers the new "contingent labour force." But this hides what it means for workers, who are being treated as the human equivalents of post-it notes: marginal add-ons used briefly, then discarded, without a sound and without leaving a trace.

People become "costlessly replaceable." They become quasi-automatons of the computer system, programmed to operate as extensions of its software and logic and to act strictly according to preprogrammed cues. Furthermore, once the wires have been laid and the jobs restructured into extensions of computer systems, the stage is set for a new global and local distribution of work: digitally dispatched to remote call-centres or to homes, trucks, cars, and garages turned into "virtual worksites" through a modem on-ramp to the Internet or information highway. Increasingly, too, more and more work can be transferred from paid employees to unpaid consumers using automated banking machines and automated gas pumps and operating the dialling end of 1-800 numbers for everything from shopping to tele-learning.

Then there's the price of admission. Would-be Internet entrepreneurs and information-service providers are already finding themselves being shouldered aside by big corporate players such as Bell and its Internet service and Microsoft with its integrated systems, applications, and networking software, plus content—all structured as one package and priced at irresistible economies of scale and scope. It could be a problem for anyone to participate just as a citizen, home-based worker, or consumer. Even before the $4-a-month increase in basic telephone service charges, 35 per cent of the people using the Daily Bread Food Bank in Toronto had no phone, because they couldn't afford it.[13] A 1994 Angus Reid poll found that 60 per cent of households with incomes above $60,000 had personal computers, compared to only 25 per cent of those with incomes between $10,000 and $30,000.

"There's going to be a technological underclass and an overclass," one of the pollsters said. "This upcoming information age is going to be dominated by the haves."[14]

The Digital Divide and the New Inequalities

The iron curtain has been torn down, it seems, only to be replaced by a silicon curtain, an invisible digital divide between the rich and the poor, the technologically enfranchised and the technologically disenfranchised. As we'll see, it also divides those who work with computers and control them from those who work for computers and are controlled by them.

In this restructuring process, the tools of communication, linked to computers, are becoming the new tools of production—as well as of distribution, marketing, and consumption. They are also enclosing us in a whole new environment both for working and for life. Furthermore, this environment is fully programmed and programmable, alive with machine intelligence and remote control. Because of their wraparound effects, the new tools of communication could also become the new instruments of political, social, and cultural domination. This is why the term "paradigm shift" is relevant here. The changes are altering fundamental patterns of social life. This is also why some very good proposals, such as redistributing purchasing power in a new social economy and cross-subsidizing new human work by taxing digital bit streams by volume, as in analyst Arthur Cordell's proposed "bit tax," are not, in the end, enough.[15] These approaches do not address the deeper issues of the shift in power, control, and even worldview occurring under the surface of restructuring.

Taken together, the restructuring of work, corporations, and national economies is launching the new economy into *cyberspace*: that is, an electronic ether-space under the helmsman-like control of systems centres and software.[16] Operating on a fully digitized base, and requiring that base for participation, this new economy is blurring the distinctions between corporations and industries, between production and consumption, and between the home and the office or store as a place of work. Both public-sector and private-sector institutions and every economic sector from resources to manufacturing and services are being turned into extensions of management information systems. The new economy is extending the scale of monopoly organization to its fullest, while ironically making it invisible through the new digital form of corporate organization called the "virtual enterprise." Control is being centralized more than ever before.

It is too early to map and fully define the political economy of cyberspace. But clearly a whole new political economy is emerging through the information highway and the restructuring that has laid the foundations of this highway—especially in the relations of power and control imbedded in it. People are not only losing control to increasingly remote information systems, but they are also coming to be controlled by them, and by the corporate consciousness accompanying them. This happens particularly as they come to be enclosed in a context and definition of work entirely programmed by computers. Software programs for computer-monitoring and performance review, buttressed by corporate cultural training and new management programs such as total quality management (TQM), are adjusting people's minds to see through the eyes of the corporate vision statement, to innovate in purely systems terms of faster and cheaper equal better, and to work at the dynamo-like pace at which the technology itself is advancing—doubling its capacity every two years.

I call this "training for compliance"[17] and a fulfilment of Frederick Taylor's original philosophy of scientific management. Once people are totally closed off inside a fully programmed work environment, once they are wired in through computer-monitoring and performance measurement, they have little choice but to comply: to willingly follow the logic of flow-through optimization, willingly participate in fine-tuning the new work model. Once restructuring has reached the point of closure, it is hard not to think as the computer thinks; it is hard to think for yourself.

In themselves, the issues of electronic surveillance and of people becoming hardwired into fully programmed operating environments raise critical questions about human rights and social justice. In their all-encompassing, totalizing effects, the issues also signal a dangerous closing down of public imagination around how to structure and govern the information highway and the political economy it will enable. They recall George Grant's bleak prophecies of the 1960s, foretelling the assimilation of society within the empire of technique and technology, and John Ralston Saul's more recent prophecies of a new corporatist society in which experts from labour as well as business and government assist in adjusting people to fulfil roles in a metasystems society instead of fulfilling themselves as citizens in the larger social community.[18]

The high-capacity multimedia technologies associated with the information highway could open the way to an inclusive, participatory knowledge society. The promise and hope of such a society is enticingly apparent in the diverse flowering of community, information, and even culture on the Internet. These networks could yet allow us all, as McLuhan hopefully envisioned, to extend our eyes, ears, voices, and consciousness around the world in a rich ecology of reciprocal interactivity; but only if these extensions are designed and structured to be truly democratic, inclusive, and participatory. It is imperative that we control the technologies by programming them directly ourselves, so they do extend our own particular minds and don't turn us into extensions of a few corporate minds, our freedom of expression reduced at best to multiple choice.

The unfolding trends are now taking us in the opposite direction. With the choices made so far, technological restructuring has served to expand and extend work for only a privileged minority. For what looks like a growing majority, it is not knowledge that the new economy wants, but dexterity: manual dexterity and mental dexterity to adjust to new operating systems and cope with the pace involved. Even graver implications are glimpsed through McLuhan's metaphor that speaks of technological extensions of human senses being accompanied by an amputation of other senses. The large-scale, remotely programmed, and controlled systems are taking us towards an amputation of our senses and intelligence, towards the closure of human communities within the all-pervasive extensions of corporate systems' eyes and ears, minds and consciousness. These systems, having come to surround us, are now threatening to reduce us to what both Grant and McLuhan called "servo-mechanisms" of technology, involuntarily or voluntarily plugged into the networked living/working environment of postindustrial society and programmed to speak only in its language of instrumental rationality—not a language of human reason, compassion, and community. Marcel Masse calls it an "acultural, alingual technology."[19] Others, extending ecofeminist Vandana Shiva's metaphor, have called it a global monoculture.

The new economy promises to be less and less *our* economy—sustaining us all with jobs, livelihoods, leisure, and opportunities to participate, and simply a chance to live at peace with ourselves, our families, friends, and neighbours—because it is not grounded in the social environment of human communities, let alone the natural environment, with a finite sense of time and space. It is grounded instead in an entirely simulated environment: that of the corporate systems economy driven by global stock and bond markets and their constant appetite for profit margins. It is as though a machine has been installed at the heart of the world: the international financial market, wired and digitized to become a perpetual-motion machine, leveraging a global division of labour, feeding off its disparities in gender, generation, race, and geography, and in turn feeding an endless drive for more productivity, more transactions, bigger and bigger scale, and faster, more integrated technology. Everybody and everything are being adjusted to become an extension of its consciousness and the operational arms of its logic, world without end.[20] It is terminal fast-forward—or perpetual "out-performance," as a mutual-fund ad put it.[21]

In the last thing he wrote before being silenced by a stroke, an essay for *Maclean's* on the dawning 1980s, McLuhan worried that the scale and pace of the new technologies might be threatening our humanity. "In the '80s, there will be a general awareness that the technology game is out of control, and that perhaps man was not intended to live at the speed of light," he began.

"Excessive speed of change isolates already-fragmented individuals and the accelerated process of adaptation takes too much vitality out of communities," he continued. "It might even be said that at the speed of light man has neither goals, objectives nor private identity. He is an item in the data bank—software only, easily forgotten—and deeply resentful."[22]

None of this is inevitable. The changes so far have been the result of choices made in the design and implementation of technology, and in organizational restructuring and policies around it—choices over which people as workers, consumers, and citizens have had no real control and little say. To challenge these changes, and to reprogram the priorities of those systems, we must first break the immobilizing gaze of the adjustment regime.

Getting Grounded in What's Happening to People

The struggle to control the technologies of the new economy must begin with a struggle to control perception—that is, the ability to think through our own sense of what's real and important, and not to think solely in terms of the official realities of technological restructuring provided by corporate and corporatist managers. If we are to articulate an alternative restructuring agenda centred in people and human communities and not in the expansionist global money markets, we must first confront the almost Orwellian perception-management accompanying the corporatist agenda, and its ability to impose the corporate-systems reality as the only one that counts.

The Globe and Mail's non-coverage of the twenty-six thousand people who stood in line through one of the coldest days in January for a chance at a decent-paying full-time job is only one small example. In that case, unemployment was framed out, while the employment represented in the UPS move, and Frank McKenna's information-highway initiatives to get it, were framed in.

There's also the selection of some people as experts and authorities, and the marginalization and silencing of others. For example, a study of Canadian Press wire stories on the economy over a twelve-month period revealed that the discussion was dominated by economists from the Fraser Institute and C.D. Howe Institute. The Fraser Institute, a right-wing think-tank, was quoted in 140 stories, while the more left-leaning National Council on Welfare and Canadian Centre for Policy Alternatives were quoted in 17 and 16 respectively.[23] Equally, government cutbacks in grants for writers, filmmakers, and video artists as well as for women's groups and other social-justice groups have cut the diversity of perspectives and generally chastened and silenced the voice of critical opposition.

For the most part though, the manufacturing of official reality doesn't require deliberate choices to include or to exclude or anything resembling a conspiracy. It operates at a systemic level, through the centrality of experts in official discourse and expert-framing institutions like *The Globe and Mail* in the distribution of that discourse, and in the centrality of objective rationality as its official language.

This is a key controlling ingredient: the language of objective rationality, with its systematic filtering out of lived experience as authoritative knowledge and the replacement of that experience by designated experts and objectified facts. This language translates people from being subjects of their own story and culture into objects.[24] They cease to be people in social and moral contexts and instead become statistics in abstract categories such as "clerical" or "production worker." They become pertinent only in terms of the discourse, which is a closed system of reality totally out of their control.

As John Ralston Saul phrased it: "Writers and their pens, having invented the Age of Reason, are now its primary prisoners, and so are unable to ask the right questions, let alone to break down the imprisoning linguistic walls of their own creation."[25]

It is also a highly privileged prison. As communications theorist James Carey argues, today's experts are willing to share their data with anyone. "What they wish to monopolize is not the data but the approved, certified authorized mode of thought, indeed the very definition of what it means to be reasonable."[26]

As I learned from my own participation, the objectifying language of the discourse also transforms writers and researchers from subjects of our own observations and reflections into instruments of the discourse, accountable to its frames and terms of reference, its rules of evidence. In coming into the fold as experts, we exchange credibility on our own terms as whole people—citizens in society—for credibility as designated spokespersons and specialists in the eyes of the discourse. We also remain credible only so far as we retain our scrupulous objectivity, checking our passions at the door. In other words, we are experts as long as we speak within the designated frames of reference and keep within the squares of the given terms of reference—what Welsh critic Raymond Williams calls "key words."

Such words have preordained significance. The discourse and those people and institutions who adjudicate it assign them their weight and authority. Words like "deficit cutting," "global competitiveness," and "productivity" are key words of our day. They anchor discussions at conferences and in the media. Repeated by designated experts (who confirm their status as experts by using them) and by public authorities, they inform public perceptions, predetermining what people will pay attention to and take seriously. And anything outside those terms and frames of reference somehow slips from the public mind. It carries no weight. So it "dematerializes" and "evaporates" as somehow less than real, as Rick Salutin put it once in a column entitled "The Picture Outside the Frame."[27]

Implacably, and in the seemingly neutral name of objectivity, the discourse on restructuring controls public perceptions of what's happening. Having first contained the discussion, the discourse then names only one reality into being within it: the reality of the global market economy. Within that framework, people become clerical or assembly-line workers: stripped of all identity and moral claim except that which relates to the skill needs (such as they are) of the new economy. In that one stroke of a term of reference, what should be a debate between two moralities—one of people and human communities, the other of corporations and corporate economies—is silenced. What should be a struggle between two logics—the logic of a global systems economy versus the logic of the living social context in which restructuring is taking place—is stilled. It is not a question of balancing and reconciling the systems-efficiency logic of corporate restructuring with the logic of society as a whole. The only context is the rational-systems one of the discourse, structured as a debate between optimists and pessimists and between alternative adjustment measures. A moral, social debate becomes a technical-management matter, all simply a question of logistics.

Participants like myself have been similarly transformed: from impassioned voices from actual communities into authorities in the discourse. Invited, as Dorothy Smith has put it, to participate in the "conceptual practices of power," we become privileged experts in the corporatist adjustment agenda, instead of daring to stay outside, in the real world of present lived experience.

Elsewhere I've likened my makeover into a designated expert speaking the key words of the discourse to someone who has mastered the fashionable techniques of flirting. I learned to walk the walk, to talk the talk. I could debate "redundancies," "downsizing," and "deskilling" with the best of them at national and international conferences, spouting figures on long-term unemployment prospects big enough for headline coverage. Now, I sense a further analogy—to the simulated language of Newspeak.

As Orwell spelled out in an appendix at the end of *1984*, Newspeak had a double purpose. First it was meant to establish a limited new set of words with built-in emotive meanings. Hence good, double-good, and double-plus good were intended to replace a whole range of traditional words such as "beautiful" and "just" and in so doing bypass the need for memory or any personally grounded meanings through which people could check out the authenticity and relevance of the words they heard and used. Then came Newspeak's second purpose: to systematically eradicate those traditional and personal meanings, to eradicate memory itself. With Newspeak, you didn't have to think for yourself; in fact you couldn't. Everything was rigidly codified. Applied here, it means that new investment and everything that contributes to it are good, double-good, or double-plus good; and whatever detracts from this—the cost of feeding the hungry, nursing the sick, caring for the young and the old—are negative burdens, coded as "bad," "double-bad," to be dealt with through spending cuts and privatization.

Today's society has little need for an overt conspiracy to corrupt language, as Orwell suggested for the world of his novel. The corruption is almost built into the system of fast-forward media communication, geared to rapid consumption, and to constant changes in fashion and other news. So today we don't hear too much about people being fired, laid off, and losing the identity they'd spent a lifetime developing. We hear of "downsizing," "re-engineering," and "reorganizing for flexibility." Jobs are "shed" as though this is as natural as the change of seasons. Jobs are "out-sourced."

People are "released" from employment, or sometimes even "dehired" and "unassigned."

The drift towards a sharply class-divided society is served up in the strictly technical terms of "two-tiers" or in the confectionary terms of a jelly doughnut, with full-time people being the jelly and the rest of us the dough.

The words used have no roots, no connection to the material structures that organize our lives. They have no history against which we can check out what's happening in the context of personal and collective memory. Like the rhetoric and mythic signifiers of advertising, they are coined and used for effect only. And like all advertising, as McLuhan warned, they can work us over.[28] As they do this, they reconstitute us as programmed objects in the metaworld of the new economy as projected on the telescreen of public perception. As living subjects, we might still feel anxious, but we are also vaguely reassured—enough to take the edge off our anger and paralyse an impulse to take a stand by naming what's going on as we see and sense it.

Janet Dassinger's experience might be typical here. She spent many days over many months in the early 1990s representing labour (specifically as Director of Training Programs and Policies at the United Food and Commercial Workers Union) in one of many tripartite sectoral committees (business, labour, and government) sponsored by the federal government to create "human-resource development strategies" (HRDS) to co-ordinate adjustment measures for restructuring in the 1990s. She tried to represent workers' concerns about job security and deskilling, but constantly found the discussion slipping under her feet and moving towards talk of "training for new technologies"—training that she doubted most workers could afford.

"You start off thinking if you can just decode it—their language—you can be strategic," she says. "But the terms are all so vague and abstract—things like 'learning outcomes'—you don't know whether it's the business agenda or the labour agenda you're talking. You spend that much time with them too, and you get these charming smooth people from the consulting firms, you begin to think in their terms. It becomes your working language. You don't say what you mean any more. It is H.R. this and H.R.D.S. that. I can't believe the amount of their language that I'm taking on. It is very seductive.

"Meanwhile, I should be at the barricades. I need to get grounded."[29]

Another difference between Orwell's imagined world and our own world of the 1990s is that Newspeak and identifying with Big Brother have lost their coercive edge. The propaganda of agitation has been exchanged for the propaganda of integration, conformity,[30] and consent. The tone is closer to that of Aldous Huxley's *Brave New World.* Now people want to be plugged in and tuned in, and they are enticed to identify with free T-shirts, microwave ovens, cruises, and cars.[31] Still, intimidating levels of unemployment add a strong undertone of coercion to the equation.

Both novels are tragedies. In the end Winston Smith was successfully (though forcibly) adjusted and came to identify completely with the machine world of Big Brother. "I love Big Brother," he said, which today might be, "I love my McJob, I love my mutual fund, and I love the IMF."

A digitally networked world could fulfil Marshall McLuhan's hopeful dream of a global village, with in-depth personal involvement for all. The global village could be a corporate systems monoculture, or it could be an extension of existing human communities in all their vital diversity. The local could be an extension of global uniformity, or the global could be an extension of local diversity.

To claim that second choice, we must reclaim our own powers of naming, and use them to articulate a new critical discourse on technology and the information highway.

2

The Chip and Programmed People: The Real World of Restructuring

The goal of a technology is incorporated a priori in its design and is non-negotiable.
– Ursula Franklin

We now move towards the position where technological progress becomes itself the sole context within which all that is other to it must attempt to be present.
– George Grant

Policy makers don't understand the real character of Microsoft yet—the sheer will-to-power that Microsoft has.
– Mitchell Kapor, computer-software entrepreneur

WHILE PUBLIC ATTENTION has been locked on deficits and social spending cuts, the ground has shifted under our feet. In a convergence of computer and communications described as the most powerful juggernaut in the history of technology, the whole basis of the economy has been digitized. Now the economy itself is moving inside the information highway, inside its network infrastructure.

The highway is becoming a worksite in its own right—a "virtual worksite" of automatic switches and processing software. It's also a virtual mall, virtual moviehouse, library, clinic, and classroom, whatever. The highway is a medium for transacting all kinds of business, individually as digital-phone button-pushers and collectively through virtual corporations and enterprises. It is a channel for the local and global distribution of work and subcontracts, and a delivery mechanism for a range of computerized goods and services—including upgrades for home and workplace appliances, patient-care management services, and corporate financial, administrative, and management systems.

The information highway is becoming our new operating environment, our "surround," as Marshall McLuhan predicted; or our new "nature," as Jean-François Lyotard put it, describing data bases. And we are all expected to adjust. If we want to stay in the loop, if we hope to survive and thrive, we have to keep up with the technology.

The social transformation is mind-boggling, and it began with something very small: the microprocessor, the "computer on a chip," invented in the early 1970s. This invention would power the computerization of almost every line of work and the integration of automated subsystems to create manufacturing and management information systems. Through the late 1980s these systems were networked into larger systems both within corporate structures and among them. Gradually, too, communications became more central, and "netware"—or networking software—became as important as hardware and traditional software. In the early 1990s the focus started to shift towards the networks themselves as the site of economic activity: not just the private corporate networks and local networks that have existed since the mid-1980s, but new generic networks and network services (and software) available through the powerful information highway infrastructure. Those networks are no longer just a means to an end, but an end in themselves. The tools of computerized communication have not just become the new tools of production. They are increasingly becoming the new tools of distribution and consumption, learning and healing, research and knowledge adjudication, and even governance. These tools have not only automated work in every sector of society, but they have also created a whole new context for that work: a machine-intelligent context of silicon work cells that can be remotely programmed, monitored, and controlled. The tools are redefining work, and the power relations involved. They are also transforming almost every institution of our society, and even society itself.

The information highway marks the coming into being of the new economy. Its global information networks and the management information systems they support are lifting the corporate economy free of time and space, free of geography and of history with all the traditional restraints, values, and social priorities. They are creating a digital "virtual" economy in cyberspace, one whose virtual presence is created through the "press enter" command and negotiated through shared file agreements or restrictive file and software access. This economy is governed by technical performance standards and global economies of scale, scope, and speed. Increasingly too it is largely answerable only to these systems priorities. Postindustrial virtual corporations will increasingly operate inside the infrastructures of the information highway, beyond the reach of democratic voters.[1] In connections made through "bytes rather than bricks," they will conduct their business, making deals to share production and markets and dispatching work to call-centres, "agile" factories, and workshops and teleworkers in homes, garages, trucks, and cars all over the country, the continent, and even the globe.

In a sense, these virtual corporations are the ultimate fulfilment of what economic historian Harold Innis called the bias of communication in the modern commercial era. The bias, as Innis observed it, has been towards fast, distance-bridging media of communications and large, monopoly-scale developments around them. The new virtual forms of corporate organization emerging in today's globally networked economy may represent the ultimate in monopoly-scale organization, in which the structures as such disappear into pure digital communications and networking capacity.

As Innis also pointed out, monopoly-scale structures are inherently unstable, for instance, in their rigid and generally escalating requirements for capital investment and maintenance. They are also inherently unjust in the way they consolidate control into fewer, more remote centres and as they extend inequalities between centre and margin in the systems they create. But, again as Innis pointed out, these structures are not deterministic, at least at the time of their initial development. They are social constructions, and their construction can be renegotiated—if they can be successfully challenged and critiqued.

The debate so far has been like the sound of one hand clapping: almost entirely one-dimensional, certainly with respect to the information highway itself. In launching the report of the Information Highway Advisory Council in September 1995, the council's chair, former McGill University principal David Johnston, told the media that "the driving force [behind the information highway] must be competition."[2] The report itself reflects this monocultural view. Throughout it repeats the call for market competition as the core principle of design and governance, while either leaving other priorities such as broad social and cultural participation in vague rhetorical limbo or taking them as achievable within the master vision of corporate competition.

The Disappearance of the Restructuring Debate

Thirty years ago, in the 1960s, there was a debate about the destructive impact that the new technologies might have on jobs, work, and even the fabric of society and the meaning of life. In the United States, physicist J. Robert Oppenheimer, who led the team that built the 1945 atomic bomb, joined a group of scientists, academics, and economists in calling for a national dialogue on automation, warning of its destructive effects. In France, Jacques Ellul published his prophetic *La Technique*, warning of a new tyranny associated with technology and its rule, which he saw enclosing the social environment and the public mind. In Canada, George Grant extended Ellul's ideas in a double lament for Canada's absorption into the U.S. commercial empire and the absorption of society within "the empire of technique," ruled solely by the "will to technique"—that is, by strictly technical values such as faster and more efficient and productive. Grant saw this as a grim fulfilment of Hegel's "universal and homogeneous state," in which there are no ideological differences because everyone thinks the same. McLuhan took a more optimistic view, predicting a spiritual, quasi-Pentecostal union through the meeting of minds and the global extension of human consciousness in cyberspace. He also counselled a personal stance of the critically aware insider.

On a more pragmatic level, Justice Samuel Freedman of Manitoba chaired a federal inquiry into technological change in the railway. His focus was on Canadian National's switch from steam locomotives to diesel, which allowed the railway to "run-through" towns that it had previously stopped at for water and coal, and where work crews had stayed overnight. Freedman's arguments went to the heart of technological restructuring as a moral and political issue. Technology, he said in his 1965 report, no longer involved simple work tools, over which management could reasonably expect to have full control. Now it involved vast infrastructures and operating systems and altered the whole context of people's lives and livelihoods. Because of technology people were potentially losing jobs and the whole basis for ever again employing knowledge and skills that they'd spent years accumulating. Not only would the "run-throughs" associated with diesel conversion destroy social relations, but they would also threaten whole social environments as CN pulled the plug on what had been a stable pillar of small-town economies.

Freedman argued: "The old concept of labour as a commodity simply will not suffice; it is at once wrong and dangerous. . . . Hence there is a responsibility upon the entrepreneur who introduces technological change to see that it is not effected at the expense of his working force. This is the human aspect of the technological challenge, and it must not be ignored." Accordingly, he wrote, the long-standing tradition of managerial control over technological change had to be replaced by negotiation—negotiation between labour and management on the basis of parity.[3]

By the late 1970s and early 1980s, when the government was sponsoring conferences and research on technological change, the moral terms of the debate had been replaced by economistic and technical arguments. Debate focused on job losses or gains and the skill needs of an economy adapting to more productive technologies. Estimates of the diffusion rate of computerization and productivity gains were taken as the critical determinants in computer models projecting job losses over time. These in turn were compared to the projected growth in new employment that would presumably emerge as increased productivity triggered increased economic growth.

Those projecting a lively compensatory job growth were called optimists, while those of us who worried that people were losing out, and that through computerization the at best precarious balance of power in our democratic society was shifting to the side of corporate management, were called pessimists or antitechnology Luddites. Still, there was an element of social justice in the debate. It was implied that if the effects looked bad enough, the government would intervene in the general interests of society. Regulations would be needed to manage restructuring, to ensure that it worked in the public interest, opening up new jobs, expanding the prospects for meaningful employment, and spreading the benefits of productivity increases as equally as possible. Meanwhile, the debate went on and on, with more and more expert reports, more and more long-range forecasts, and with the bureaucrats organizing these debates continuing to declare uncertainty over the "net" long-range effects of restructuring, and calling for more research.

When the government changed with the election of the Progressive Conservatives in 1984, the debate itself was restructured. The Mulroney government's first major policy document, *A New Direction for Canada* (November 1984), defined a lack of competitiveness and a poor investment climate (blamed on the cost of complying with labour standards and other regulations) as the problem and declared technological innovation and deregulation as the solutions. "Canada's future economic performance will in large part be determined by how successfully Canadians respond to the challenge of an ever-quickening pace of technological change and an increasingly competitive world trading environment.... If we are to be competitive at home and abroad, we cannot impede change."[4]

What had been at least partially a debate about conflicting values and priorities around technology was replaced by a technocratic prescription: a set of measures for adjusting to technological restructuring. Furthermore, these measures were based on the terms set out by globalizing business, not by the local, regional, or even national business community. As if to assist this shift, at least two critical forums in Canadian public-policy debate—the Science Council and the Economic Council—were simply shut down.

What was left of the restructuring debate was then overtaken by a new discussion, about free trade. Once again experts came to the fore, and once again debate became a numbers game. This time it was about job losses or gains due to free trade, with a side debate about the price of chicken and the cost of living in general. The deal was masterfully marketed through a multimillion-dollar corporate ad campaign and untold millions' worth of government publicity, all keyed to what has been called the "no-brainer" phrase of "free" "trade." The sizzle of "free" sold people on the new "real" world of free-for-all corporate expansionism. And the meat of the trade document itself—like the steak for the dog in the burglar's bag of tricks—distracted us from what was really going on.

By the time the evidence of plant closures and the shift of manufacturing (and some information services) to cheap-labour countries in a new continental division of labour came in, free trade had been eclipsed by a new issue: the *deficit* and *debt reduction.* In the 1990s, eliminating the deficit by cutting public-sector spending became governments' top priority. Cutting public-sector employment and public services, especially to the poor and the powerless, became the most respectable thing governments could do.

There was clear evidence that much of the debt had resulted from interest charges related to previous deficits. What's more, a large portion of the deficit was caused by the drop in income-tax revenues brought about by increased unemployment and underemployment, combined with the strain this placed on social assistance programs. Author Linda McQuaig argues that $20 billion of the $49 billion deficit of the early 1990s was attributable to lost taxes because of unemployment.[5] Economist Diane Bellemare tallied the double costs of the new high levels of unemployment—lost tax revenues plus social safety nets—and arrived at a figure almost equivalent to the deficit.[6] As Ramona Lumpkin of the University of Windsor told the House of Commons Standing Committee on Human Resources Development: "Social security programs are seen as a drain on the economy. . . . But it is equally true that the national economic situation has created a massive drain on social programs."[7] Finally, deficit spending had been respectable economic policy for fifty years and continues to be practised by private enterprise—to the tune of $2.2 trillion in the U.S. corporate sector by the early 1990s.[8]

Despite all this, the 1995 federal budget and its enabling legislation began the process of ending federal spending on health, education, and other social programs and of laying off forty-five thousand public servants. The $7 billion loss in social transfers for education and health was expected to increase postsecondary tuition fees by 300 per cent. It could bankrupt the health-care system, as it forces more hospitals and clinics to shut down and others to scale back services.

As well, under the Newspeak rhetoric of "social-welfare reform," the Liberals continued the Conservative government's scaling down of unemployment insurance (UI) by lengthening the work periods required to qualify, shortening the coverage period, and cutting the UI budget by $8.5 billion by 1999.

Just as a lot of the talk about free trade didn't make sense in terms of how trade in goods and services really works, the cutbacks don't make sense in the context of public debt and the real sources of the problem. But they do make a lot of sense in the context of technological restructuring.

Opening the Black Box of Technology

First, a word or two about technology itself, and the black-box mentality that prevents many of us from coming to grips with it. Some people assume that technology is a neutral tool that anyone can control, and so they are unprepared for (and reluctant to acknowledge) how its larger construction can limit choices and exclude people from the decision-making process altogether. Others assume that technology is deterministic, that it is a force like nature ("creative gales of destruction" being a popular phrase here) and so big and powerful that only huge institutions like government can grapple with it, and even they can do little more than mitigate its effects.

But technology is a social construction. Its design, organization, and use reflect the values and priorities of the people who control it in all its phases, from design to end use. After the design has been implemented, the system organized, and the infrastructures put in place, the technology then becomes deterministic, imposing the values and biases built into it.

U.S. civil engineer Robert Moses built his racism into the construction of overpasses on the New York freeway out to Jones Beach on Long Island by designing structures so low that city buses, many of which would carry African-Americans from Harlem, couldn't get through. According to his biographer, this exclusion wasn't accidental. It was part of Moses' design.[9]

Christopher Sholes's QWERTY keyboard design earned him the title of "father" of the typewriter. The design, however, was geared to preventing keys from jamming in the carriage, not to helping typists at their work. It put machine efficiency over human efficiency. By the 1930s a later effort, the Dvorak design, had been demonstrated to improve typists' speed and reduce typing errors. By then, however, an *infrastructure* of factories tooled to make the Sholes model, typing schools dedicated to teaching it, and millions of typists already trained in it all prevented the new model from taking over.[10]

Clearly, infrastructure—the sometimes forgotten backdrop to technological tools—is critical. It can limit choices in how tools can be used; for instance, through pricing policies, access routines, and operating speeds or delays. It can also often determine what tools will become available in the first place and restrict freedom of choice to multiple choices—such as programs on cable television—provided by mass-distribution conglomerates.

Values such as inclusive participation and reciprocity versus restrictive participation and centralized control are built into the infrastructure over time. Once these values are built in, as is happening now through choices made in technological restructuring and in the information highway, they will strongly influence the nature of the economy and of society. The bias in building the systems and infrastructures for the new economy has largely favoured the centralization of control into a few hands, as it has in past infrastructures of communications.

The history of the telephone provides instructive lessons in this. For example, in the early years of telephony, inventors came up with a switching device that users could install in their telephone boxes to protect their privacy on multiuser or party lines. The device could have opened a path for greater user control of the phone; for instance, users could have arranged their own conference calls and even used the phone for quasi-broadcasting.[11] But the telephone companies cut off this path by establishing a centralized system in which the telephone company controlled all the switching using individualized single-user lines. This was by far the more expensive technological choice; but it delivered on the values of centralized corporate control (including the possibilities for greater income through switching transactions), plus restrictive end-user participation.

In an interesting parallel to the current early stages in the development of global networking, telephone operators were also important innovators in developing and popularizing the technology. Often the sole representative of the telephone company in a community, they had a fairly free hand in putting the new technology to use, and they were largely responsible for its growth as a domestic tool of communication and a medium of community culture. Much as Schoolnet, community non-profit freenets, and the more global and commercial multimedia worldwide web are creating cultural and social networking services through the Internet and the information highway, early switchboard operators created an all-purpose community information service as they patched people through to doctors' offices and the fire hall, gave out the time, and passed along messages and even broadcast community announcements. Indeed, the operators' reliability as communicators compensated for the technical unreliability of the early telephones and prompted customers to subscribe to the service.[12] Later, after the system had matured, operators became more and more restricted to narrow job functions, and in recent decades more and more of their jobs have been automated, standardized, and thoroughly regimented. The development path of communications as culture and community was all but eclipsed as the corporate infrastructures asserted a more strictly commercial model of communications as transactions.

The short but eventful history of Microsoft offers another lesson—here, in the rigidities that can accumulate through centralized corporate control. By the early 1990s, Microsoft's computer-operating systems had gained 80 per cent of the computer market, and its applications software had become the industry standard in a host of user functions. Then the 1995 launching of Windows '95 systems and applications software package pushed Microsoft's networking and Internet software as a de facto industry standard into the third networking phase of computerization, consolidating a global networking infrastructure under this one corporation's control. As an industry insider put it, "It's a clear extrapolation of their operating system monopoly into the network application market."[13]

Powerful as the infrastructures are, there is still considerable power through public regulation and in the day-to-day context of technological practice in which people do their work and take care of business. Through their own personal knowledge of the work process and their own human sense of purpose in it, people have considerable power to renegotiate end uses of technology, to finesse and fiddle ways around things like bugs in the system—and even to redesign and restructure the larger systems and supporting infrastructure. So even as we seem to be approaching closure inside the information highway, even when people are being integrated as operating units of cybernetic corporate economic and service systems and becoming adjusted to thinking through the corporate mind-set, there are still openings. We still have the time and opportunity to gain control.

But the stakes are high. In the bulk of technological restructuring since the 1960s, the choices have largely favoured exclusive, restrictive participation and centralized control rather than inclusive participation and local control and reciprocity. At every level, from design to local restructuring and systems management, the choices have overwhelmingly advanced what scientist and activist Ursula Franklin has called the prescriptive production model over the holistic growth model.[14] To reverse this state of affairs, we must first understand how we have reached it.

Chicken Barns and Push Buttons: The Story of Restructuring So Far

With the microchip's invention, automation brought computer intelligence to every conceivable line of industrial and bureaucratic activity, letting more and more machines run themselves everywhere from chicken barns and cattle feedlots to assembly lines and offices, hospitals, stores, and banks. Robots replaced human welders, sensors replaced human touch and smell, and optical scanners replaced eyes and hands.

In the service sector—the place where most new employment has been created in the postwar period, computer systems and software have replaced human filing, collating, reporting, administration, and, increasingly, analysis, evaluation, and management as well. With the microchip built into communications lines too, information-processing and automated-production equipment could be integrated into larger and larger systems and related management networks, which in turn allowed for faster decision-making and more centralized control while simultaneously facilitating more decentralization of work.

In the 1980s this trend yielded a tremendous increase in productivity and also set the stage for organizational restructuring, in particular a hollowing-out of clerical, administrative, and middle management work, with production work integrated into almost cybernetic modules and workforces reduced and drastically realigned. Along with all this came a more global division of labour and, locally, a polarization of employment into full-time jobs on one side and jobs of precarious status on the other, as well as jobs polarized by hours of work and pay.

The current networking phase, including a frenzied "ramping up" to the information highway and a proliferation of Websites on the Internet, augurs another wave of organizational restructuring for the late 1990s, as more corporations enter the political economy of cyberspace, with "virtual" forms of corporate organization and the organization and management of work. This wave promises to destroy even more work, in every economic sector. It will also hollow out more of the middle ground of work, not just within corporations but between them too: instead of going through local suppliers, wholesalers, and warehousers, corporations will bypass them, the way Wal-Mart now does.[15] Increasingly, too, corporations will sell to consumers directly, via 1-800 numbers and websites, bypassing local stores and distributors. This shift could devastate small and medium-sized businesses and local and regional economies.

The evidence of various case studies from the 1980s and early 1990s (to be discussed in more detail in the next two chapters), documents the links between technological restructuring and the drastic realignment of employment. These trends are:

- *the collapse of work itself*: through closures, layoffs (sometimes attributed to recessionary economic times), and job vacancies going unfilled.

- *underemployment*: consistently, in almost every industry and occupational group, full-time workers have been replaced by part-time workers, full-time jobs have been re-engineered into part-time or contract jobs, and job expansion has been concentrated in positions offering less than thirty hours a week, and sometimes less than twenty.
- *jobless growth*: caused by either dramatically increased output or throughput with automated systems, or the creation of new goods or services that can be both produced and delivered by software or automated systems. The privatization of computer-simplified work into the hands of consumers (what Alvin Toffler dubs "prosumers") is a major part of this trend, with effects ranging from the new permanently high levels of unemployment and underemployment to a new docility in the labour force whereby people feel grateful for whatever job and wage they can get in the face of the growing "reserve army" of unemployed.
- *deskilling and the declining middle*: although computerization has increased the need for some highly skilled, highly specialized people—in other words, it has had some "upskilling" and "reskilling" effects—in the majority of cases it has had the opposite effect. In some cases, it has sucked out people's knowledge, skill, and judgement, substituting software and "expert" systems. In other cases it has combined the tag-ends of computerized, computer-simplified jobs into multitasked jobs. Some observers cite these as examples of multiskilling, but the evidence of the actual cases suggests otherwise. Taken together, these features are consolidating the new digital divide between those who work with computers, often full-time, and those who work *for* them—often in post-it-note status.
- *the de-institutionalization of workers and work*: for most of the 1980s this issue was discussed largely in terms of the increasing trend to replace full-time workers with part-time, temporary, term, and contract workers. At the same time, work has been transferred out of bargaining units into engineering and executive offices—a harbinger of the de-unionization that is accompanying the de-institutionalization and individualization of work.
- *computer-monitoring*: although often discussed as a human-rights issue—the invasion of people's privacy—this development also signals the expanding brave new world of McJobs,

which, in turn, are linked to the computer-integration and computer-reorganization of work. Computer-monitoring also heralds a new cybernetics of labour—that is, a realm in which labour-management relations are entirely defined by the computer.[16] It raises fundamental questions of human identity and social justice.

- *credentialism*: the shift from experience-based knowledge and on-the-job learning to formally accredited learning reflects an increasing emphasis on task-based and technology-specific knowledge, and a bottom-line fine-tuning of "manpower" to the exact operating needs of a workplace. Like computer-monitoring, though, credentialism also signals a stage in the technological restructuring process. It marks a shift from a largely social context and infrastructure of work to a technological context and infrastructure in almost every line of work and industry, where human requirements are increasingly defined in terms of technical functions.

Credentialism also augers a deepening of social inequalities because of inequalities in the distribution of training hours and dollars. One study found that the bulk of corporate training, in time and dollars spent, is being concentrated in the core of full-time employees, and the more highly paid ones at that. "Given the established link between workplace training and subsequent wage differentials, this can be expected to increase overall wage differentials," the authors noted.[17] This trend solidifies and deepens class divisions, but it has a sexist dimension too. In the struggle for meaningful training opportunities, women are losing out to men. In the new competitiveness for scarce jobs, increasingly pricey credentials become a sieving mechanism for short lists, and a potentially discriminatory hiring gate.

A number of larger, macro studies corroborate the case-study trends. In general, the proportion of people earning middle-class incomes dropped by over 9 per cent between 1973 and 1989.[18] A 1988 Statistics Canada study documented a massive hollowing-out of employment in the middle ranks of wage levels, everywhere from processing and manufacturing to service industries, and a dramatic polarization of net employment growth between 1981 and 1986.[19] An update on the numbers to 1989 reflected a slight recovery from the 1980s recession, but still showed a distinct polarization of employment growth by wage levels, within both manufacturing and services and within occupations.[20] In processing and manufacturing, nearly eighty thousand jobs disappeared through the 1980s at wages ranging from $6.77 to $10.43 an hour. The bulk of the new manufacturing job growth occurred at wages above that level, although over ten thousand new jobs were created at less than $6.76 an hour. The greatest hollowing-out emerged in clerical, sales, and service occupations and in the industries of distributive and consumer services, areas in which women are highly concentrated—and where women had begun to move into the middle ground of decently paid jobs involving a degree of autonomy and personal judgement.

A 1993 Statistics Canada study added an age dimension, documenting a massive downward shift in earnings from 1981 to 1988 among people under twenty-four years of age, as well as among those between twenty-five and thirty.[21] The decline was more pronounced for men—an 18 per cent drop for the ages from seventeen to twenty-four—than for women, largely because women didn't have so far to fall in the first place because of gendered wage inequalities and the historically high concentration of women in part-time temporary work.

Men with university degrees took a deeper plunge than college graduates or even those with only grade-school education. For male university graduates, earnings fell by over 5 per cent between 1981 and 1988. Men with postsecondary diplomas took a drop of over 3 per cent. For women, the situation was different. Their earnings fell at every educational level, except the top. Women with university degrees—particularly those in the natural sciences—increased their earnings by nearly 3 per cent. However, when their education/age levels in 1989 were adjusted to the equivalent levels of women in 1981, the comparable earnings showed a drop.

These findings corroborate studies arguing that opportunities for employment—that is, for meaningful, challenging work—are not keeping pace with young people's levels of education and skill. In a survey by the Ontario Institute for Studies in Education, 40 per cent of the respondents said that people are generally overqualified for their jobs. In short, they "are underemployed," as David Livingstone, the report's prime author, put it.[22] When Chrysler Canada hired a thousand new workers in 1994, almost 30 per cent of them had university degrees, although this had nothing to do with the job requirements, which required little more than high school.[23] Finally, another study suggests that while the working population already has a lot of computer skills, most people aren't getting a chance to tap those skills; they're doing little more than rote data-processing or word-processing.[24] The evidence shreds the liberal-democratic belief, and dictum, that if you get a good education and are willing to work hard, you will find a good job and move ahead financially. The educated might be winning a job slot in the great jobs lottery of the 1990s; but they are not being employed nearly to the full extent of their knowledge, skills, and desires for involvement.

The 1993 Statistics Canada report suggested that "technological change" or some form of deskilling might explain this shift in earnings. But it lacked the necessary data to explore a possible link between technological restructuring and the deepening polarization of work hours and earnings.

Work on the fringes—part-time, temporary, term-contract, self-employment, and other "contingent" work—represented the bulk of new job growth through the 1980s. Furthermore, a 1988 study found that people in these "non-standard jobs" earned less than half of what those in standard full-time jobs earned, and few had dental plans or other benefits.[25] And most part-time workers don't qualify for unemployment insurance—although this could change under amendments to the UI legislation introduced to Parliament in late 1995.

A follow-up study on earnings and working-hours inequality documented deepening generational as well as gender disparities.[26] Younger men were working fewer hours, and more were ending up in part-time and other contingent work arrangements. Older men (over forty-five) were working longer hours in regular full-time employment—particularly in management, sales, and manufacturing. In these three occupations, the fraction of men working fifty hours or more had risen by six points for managers and three points each for the others.

According to the study, women's earnings evened out, but this was because more women were working more part-time hours, while only some women—generally in the natural and social sciences—were working more full-time hours. Between 1983 and 1989, weekly hours worked by women in part-time jobs rose by nearly 5 per cent, while women's hours in full-time jobs rose by less than half a percentage point.

The trends signal both a startling impoverishment of young men and renewed gender inequalities, confirmed in a 2 per cent widening of the wage gap between men and women in 1994.[27] More and more women are being ghettoized in the new permanent part-time workforce, with few if any benefits, including training, and with a skills, credentials, and silicon ceiling preventing their mobility. They are also part of the growing trend towards home-based work, particularly in the female ghettos of clerical, sales, and service. Coupled with the increasing privatization of health care, leaving it to a greater extent in the hands of family members—which means women in 70 per cent of cases—it seems that women are being redomesticated into the home. The private sphere could once again become women's sphere, while men reclaim the public sphere, including the voluntary or involuntary overtime that is largely worked by men.

Rigid inequalities are being built into the social environment, along multiple dimensions. For instance, a 1993 Industry Canada report warned that as a result of technologically induced job polarization, an overall "deskilling" is becoming a permanent feature of the labour force.[28] More important, the Conference Board of Canada notes a new unhinging of wage increases from productivity and profit trends. While corporations are showing record profits and productivity gains, workers' real wages (adjusted for inflation) are falling and expected to fall even further. This unprecedented phenomenon—which defies conventional economic theory linking productivity, profits, and wage increases—is thought to have resulted from restructuring. According to a Conference Board official, "structural changes in the economy" have severed the link between profits, productivity, and income growth.[29]

This is a stunning development. It suggests that the distribution of power in society is fundamentally shifting in the move from the industrial to the postindustrial economy. Where people were once generally central to the industrial and bureaucratic process and able to leverage their share of its benefits, now machines and machine intelligence are central. Where hand tools and machines requiring attentive, intelligent, and even skilled workers to operate them had been the means of production in the industrial era, cybernetic systems and intelligent communications networks have taken over, and these systems require fewer people—and less of their intelligence and involvement. A broad and inclusive division of labour, plus markets, supplied much of the wealth of nations in Adam Smith's day. Now, as analyst Arthur Cordell notes, "The new wealth of nations is to be found in the trillions of digital bits of information pulsing through global networks."[30] People have become peripheral, and their influence is being eclipsed. This in turn has a bearing on the collapsing welfare state and the collapsing income-tax base for financing social programs—especially since 1984, when the tax burden in Canada was shifted away from corporations and more to individuals.

Technological restructuring, then, goes far beyond job numbers and employment as such. Technological restructuring is a political and even a cultural issue and raises fundamental questions about whose priorities are paramount versus whose are expendable or irrelevant. In other words, why has trickle-down economics trickled right out for people? It also raises questions about who is in power and control versus who is losing power and control and even coming to be controlled.

Politically the most important change has not just been the hollowing-out of the middle ranks of managers, administrators, and professionals, which has severely reduced the middle class. It has been the replacement of human management and administration with systems software, bringing more and more work under the control of machines rather than people.

A second related development has been the restructuring of computerized work to the point at which it can be almost completely defined and managed—and monitored—by computer systems. People then find themselves working directly for computer systems, as extensions of their operating software, with no opportunity for advancement or involvement past the silicon curtain descending inside the operating system. This is the essence of the McJob and of the new cybernetics of labour: the system software controls and defines the work to be done; people are reduced to being functionaries of the system. They follow instructions and cues provided by the computer, and in return the computer gives them a read-out on their productivity, which is taken to equal their performance. Many jobs that previously involved a good deal of talent, intelligence, and commitment—for instance, in hospitals, stores, insurance and law offices, and factories—have been turned into McJobs. As the context for their work is digitized, people are being systematically stripped of their capacity for human involvement and judgement. Machine intelligence and logic take over.

But the systems don't just define what people do. Increasingly, they enclose people in an entirely programmed working environment—one so complete that the workers involved can no longer conceive of the work to be done in terms other than those provided by the computer system. Once closure has happened, they can become the technology's *servo-mechanisms*: programmed like a thermostat—servo-mechanism of a heating system—to switch on at a certain preprogrammed cue and to switch off after accomplishing a preprogrammed task.[31] This adjustment to respond automatically to set cues and instructions occurs particularly if cultural training cements their identification with the computer's and the company's goals of productivity and competitiveness.

Once the initial decision has been made to replace employees with machines and to use those systems to diminish and control what the remaining workers do, rather than to enhance and expand their jobs, further choices become biased by the relative weight of investment in systems versus people. It becomes easy, even cost-effective, to replace full-time staff with part-time, term, and contract workers and to rely more on computer software for intelligence and administrative support. It is also easier to relocate the work to take advantage of Third World wages or captive labour pools in homes or depressed regions. The current networking phase of technological restructuring has set the stage for the redeployment of computer-defined work to virtually any of the plugged-in sites along the networks of the information highway. Instead of universal connectivity as a rich knowledge and cultural network, it could mean a newly de-institutionalized, de-unionized workforce employed in virtual workshops or call-centres, or plugged in as teleworkers in their homes, with possibly rented computers and modems enclosing them into isolated silicon work cells. Furthermore, these workers can be kept under the all-seeing eye of remote management information systems, the gaze of which can sweep the information highway faster than photo radar.

On a larger scale of political economy, restructuring has enabled a re-engineering of regional and national economies into modular components of an increasingly globalized corporate economy controlled by transnational management information systems. Since the mid-1980s Canadian governments have moved this trend along under the banner of free trade, deficit reduction, and spending cuts.

Free Trade, Deficit-Cutting, and Other Structural Adjustments

More than anything, it was the timing of the Canada–U.S. Free Trade Agreement (FTA)that fit the restructuring agenda. The integration phase of computerization was under way; the networking phase was about to begin. The grounded and fairly rigid regional and national economic units associated with the old industrial era (what Alvin Toffler called the "second wave") were about to be replaced by more fluid continental and global units associated with the third wave of computer communications. The Canadian "miniature replica" model of branch-plant industrialization—in which Canadian assembly plants produced a range of largely consumer, industrial goods solely for the domestic market—had outlived its usefulness. Its dumb, electrically powered (not computer powered) mechanical technology—much of it older, second-hand equipment imported from the U.S. parent company—was obsolete.[32]

Free trade furnished the excuse to abandon this old production model, and the factories operating under it, without having to call it technological change. The trade deal's provisions allowing for "the right of establishment" and of "national treatment" by U.S. companies operating in Canada smoothed the regulatory path for continental integration as companies retooled their plants with the new postindustrial generation of computerized equipment, equipped both to function and to communicate instantly and often automatically across national borders to remote head offices. All the political and moral issues associated with abandoning Canada's original National Policy of the 1870s disappeared beneath the distracting double-plus-good rhetoric of "free" combined with "trade."

Its significance as both industrial restructuring and a new continental division of the postindustrial labour force became clear later on. Between January 1989, when FTA went into effect, and early 1994, there were nearly seven hundred permanent plant closures in Ontario alone.[33] A study of the Ontario plant closures found that slightly under half were foreign-owned branch-plants.[34] In addition, other plants relocated south, the majority of them moving to the low-wage states of the southern United States or to Mexico, where wages dropped to an average of only U.S.$2.35 an hour by 1992, 68 per cent of what they'd been in 1980. In the export-zone *maquiladores*, wages were only U.S.$1.15 an hour.[35]

Canada has gained some new investment, in the auto and aerospace industries and other high-tech industries, but for the most part the flow has been in reverse: between 1989 and 1993 there was a net outflow of private equity investment totalling $13.6 billion.[36] "Whether you like global free trade or not, the fact is, it is being driven by global investment that is [highly] mobile," said the Liberal government's international trade minister, Roy McLaren.[37]

Bruce Campbell, an author and Executive Director of the Canadian Centre for Policy Alternatives, puts it somewhat differently: "FTA/NAFTA is an economic constitution, locking in a de-regulated trade and investment regime with no agreed-upon social and environmental rules within which competition must take place."[38]

The agreement is not as binding on the United States as it is on Canada. Political economist Stephen Clarkson has described free trade as asymmetrical internationalism, arguing that in the unfolding stages of so-called trade liberalization, "The U.S. will retain the bulk of its sovereignty while its growing band of junior partners are strictly bound by NAFTA and, later, WHEFTA (Western Hemispheric Free Trade) into a new post-national dependency in which one of their governments' chief roles will be to enforce U.S. policies on their territories."[39]

If free trade makes the most sense when placed in the context of restructuring, the social-spending cuts make even more sense in the context of globalization. The new global economy seeks single or readily interchangeable operating standards—and this applies to more than just the way machines and technical calculations work around the world. It also applies to what are called "human resources" and the relationships associated with the cost of unemployment insurance and other national social benefits. The restructuring brought in under the name of free trade plus the gutting of social supports under the guise of deficit-cutting is engineering the type of low, common standard that transnational corporations desire in an interchangeable continental workforce. For example, the longer qualifying period for unemployment insurance (sixteen to twenty weeks at thirty-five hours a week, compared to six to seven weeks) conveniently adjusts Canadian standards down to the minimal level of the U.S. system. The move also levels the regulatory field for continental business practices by flattening out what corporations call "payroll taxes" on both sides of the border.

The cutbacks in funding for postsecondary education chillingly complement the restructuring agenda too. Except for a few, the children of the postwar middle class are seemingly being adjusted into McJob-like status, pushed into working as little more than extensions of computerized production and service systems and related contract support staff, if they can get work at all. They will need computer literacy and technical systems skills, including some high-level computer language skills to knit software modules together, but little more than that. Facing a doubling of tuition fees and few decent job prospects, yet "paralysed" by debt, university students talk of a new "disincentive to go to school."[40] Tellingly, university enrolments began dropping in the fall of 1995.

Workfare (augmented with cuts in welfare rates) is perhaps the most cynical social-reform item on the restructuring agenda. In the absence of real employment opportunities, its rhetoric of "a hand-up rather than a hand-out" is a cruel lie imposed by the privileged and powerful on those with no power to resist or even to repudiate the words. It also threatens to introduce a new third tier of forced labour, and to normalize a new underclass of marginal workers, as happened under the nineteenth-century poor laws. In New Brunswick, where the last pauper auction in Canada was held in 1898, paupers were contracted out to the lowest bidder—that is, to the would-be employer who promised to charge the county the least amount of money to maintain a given pauper for a year in return for hard labour.[41]

There is also the very real danger that social-service agencies could be retooled as postmodern versions of the late-eighteenth-century's "poverty police." Instead of sympathizers working in solidarity with the poor to eradicate the injustices that perpetuate poverty, well-meaning people and community agencies could become welfare recipients' keepers. With cutbacks in municipal and community services leaving a lot of necessary work undone, the agencies could almost be forced to take on workfare "volunteers" and to embrace new roles keeping these people busy and productive and under benign surveillance. By November 1995 this had started to happen. In Ontario the Kiwanis and Rotary clubs were included in plans for the "Ontario Works" workfare program, without having even been consulted ahead of time.

The cutbacks in public services fit the restructuring agenda in another sense as well. Computers, software, and networked services are filling the gaps created by staff shortages and the shift from well-experienced full-time staff to less knowledgeable part-time help. This has been going on since the 1980s, with devastating effects. One less obvious effect is that these institutions have been brought to the point of embracing more and more of the computerized goods and services offered by transnational corporations, increasingly via the information highway, at continental and even global economies of scale. If more and more services are handed over or contracted out to these corporate systems and service providers, the public-sector institutions will be effectively privatized, and colonized, from within.

Cutbacks have become an all-purpose excuse for getting rid of people, much like "recession" served the same purpose in the early 1980s and 1990s. With the funding cuts to non-governmental organizations, social-activist groups, and artists, unions have become one of the only institutions that can financially support research and critical analysis. Taken together, the changes occurring under the banner of deficit-cutting also represent a social-policy sequel to free trade, and a mopping-up phase in the restructuring agenda. Not only are Canadian social standards being adjusted downward to the charity-line levels found south of the border.[42] Not only is a national social standard being replaced by a continental one set by the United States. The Canadian welfare state, the bedrock of Canadian liberalism, is being abandoned.

Political scientist Antonia Maioni argues that the social-welfare reforms being introduced in the 1990s are replacing universality with a means test and shifting the state from the role of social protector to mere technical regulator. They also shift the basis of the social contract from the collectivity of society as a whole to the individual.[43] This, she argues, represents the assimilation of Canadian liberalism into U.S. economic liberalism and a repudiation of social activism by the federal government. The neo-liberal and neo-conservative agendas are one and the same. Both would abandon responsibility for the welfare of the country and the common good and leave everything to the market. They would reduce social relations to property, consumer, and labour-management relations and forget about relations between people as citizens in a democratic society.

Public governance and regulation are being replaced by market, corporate, and corporatist regulation in everything from communication and information highway policy to foreign affairs. A joint Senate-House of Commons report on Canadian foreign policy began by stating: "Globalization is erasing time and space, making borders porous, and encouraging continental integration. In the process national sovereignty is being reshaped and the power of national governments to control events, reduced."[44] The report went on to accept as given the new "common governance" by transnational corporations and suggested a greater role for "Team Canada" in Canadian foreign policy. "Team Canada" was the catchy trademark adopted by the two to three hundred senior executives from major Canadian and branch-plant corporations who spent months and millions on three highly publicized government trade missions led by Prime Minister Jean Chrétien in the mid-1990s. When he arrived in Vietnam Chrétien enthused: "This is the day for making business deals."[45]

While the prime minister alternated between compliant international corporate cheerleader and fumbling national leader, a new international level of corporatist governance was taking shape. An interesting hint of what this might mean for the future emerged from the 1995 G7 Summit. In an effort to "calm down" financial markets, the summit discussed plans for a new International Monetary Fund "surveillance" service: countries will be required to provide economic data that the IMF will scrutinize according to benchmarks for economic performance and stability.[46] What this suggests is that the IMF's ad hoc restructuring measures imposed on various Third World countries in response to the Third World debt crisis could be normalized into a permanent fiscal supervisory force for other countries as well. Economist Tom Naylor refers to the dealings of the IMF and institutionalized central bankers' meetings as a new "invisible government."[47]

This could be our brave new world, our future new reality. But this, again, is just a *possible* future unfolding. It will unfold that way only if we can't come to terms with restructuring as a social and human issue, if we can't shake off the spell of compliance or move away from the corporate-systems agenda—if we can't break free of the dynamo driving even the putative "winners" of the restructuring process.

Time, Moore's Law, and the Technological Dynamo

At Corel, an Ottawa firm making software applications for Microsoft Windows, the young (generally male) engineers and their managers put in up to three months of twenty-hour days to meet product-delivery deadlines.

"No one can create the No. 1 package without sacrifice. You can't do it by working 9 to 5," Corel's chief of engineering, Ed Eid, told a reporter. "You look out [the window] and it's like another universe. It's hard to imagine there's another world going on out there.

"You wonder if you're at fault for some family problems here. Most of the disputes I had with my wife over the last few months were because of work. It must be the same for others."[48]

The workplace had become the site of "stress-related breakdowns," of people shouting at each other. According to Eid, when a flu bug hit it spread like brush fire, people were so burned out.

More and more people are being driven, faster and faster, by the technological dynamo—that is, the metaphorical equivalent of a central steam engine driving everything else. Computerized filing and analytical software aids have removed the simple tasks through which people normally paced themselves with slack and waiting time, leaving them instead in a state of non-stop peak performance, with scarcely a moment unbooked. Constant change leaves people without the time to get comfortable with a new technology or system. As a librarian told me, one day you're in training, and the next day you're the expert. Other technologies, such as voice mail, fax, and E-mail, compress time. A policy analyst described them as "wild horses." You think you're driving all these gadgets, but they're driving you.

These technologies are in turn being driven by a machine law, the law of the microchip, called Moore's law. In the early 1970s Dr. Gordon Moore of Intel Corp. was preparing a speech on the future of memory chips, and came up with a "law." He plotted the capacity of past generations of such chips, then extrapolated to predict that the complexity and capacity of memory chips would double every two years.[49] And he was right.

Moore's prediction has not just become law; it has come to be enforced as law. Moore's law exerts a systemic bias towards more and more computerization to fill up greater and greater memory capacity. Everything becomes a target for possible computerization, and this direction is assumed to be the efficient and cost-effective one—double-plus good all around. With irresistible cost-benefit testimonials, the technology drives its wedge through every sector of the economy, transforming work from holistic processes governed and controlled by the people involved into increasingly standardized prescriptive processes, the doing of which is reduced to computer-defined McJobs on the ragged edge of poverty and banality.

There seems to be no stopping the expansionism of the chip and the companies dedicated to advancing it. The memory-capacity promise must be fulfilled, and no unused capacity can be left over at the end of two years, when the next generation of chips comes on the market, offering still more capacity that in turn must be filled in another two years' time, with new and better lines of Microsoft software and so on. All of which builds dependency on the next generation to come along, as the doing and managing of more and more activities from the trivial to the life-and-death are invested into computer systems. And all of this, in turn, extends the scope of what more can be computerized at more cost-effective economies of scale.

In the 1970s and 1980s, the focus was goods and services in the private sector. Now the public sector is being drawn in, taken over, and digitally transformed by these systems: in education, through "virtual classrooms" plugged into remote data bases through the information highway and computer-driven training packages offered at almost giveaway rates by vertically integrated multiproduct, multiservice information companies. The same thing is happening in health care, through "virtual surgery," remote diagnostics, and computer-driven patient-care management systems. Computerization not only spreads wider, it digs deeper and deeper, endlessly shifting work from the creative and relatively fluid arena of human relations and groups into the orbit of computer and network control. The industry literature refers to the process as "shifting cognitive loads from people to machines," as though this were a liberating thing, not a move towards total human alienation and closure within the machines of a fully realized technological society.[50]

Think again about the ads for new technology that depict computer power at our fingertips. The real finger on the button could belong to the computer systems, with their priorities for exponential expansion, and they could have *us* at their fingertips. Computerization and global restructuring could be driving us with their logic of faster, cheaper, better. They could be "speaking us" into existence in the same way as having the latest brand of beer speaks us into being in the advertising landscape. McLuhan warned about the business executive becoming the "servo-mechanism" of his watch, closed into the scheduled extension of his existence and compelled to serve that regime. Today that watch is embedded in the digitized landscapes of our lives: the systems of E-mail, voice mail, call-waiting, cell phones, pagers, and automatic production scheduling which surround us in our work and daily lives. This watch is no longer strapped to our wrists—we are metaphorically strapped to it.

McLuhan spoke hopefully of technologies as extensions of man and of human consciousness. But he also emphasized the dangers associated with the amputation of our humanity if we become mere extensions of technological systems. We could become "the sex organs of the machine world," important solely for the purpose of propagating it; "enabling it to fecundate and to evolve ever new forms"—that is, more and better machines.[51] Ask yourself: does it make any real human sense to keep doing things faster and faster, to be always trying to beat or at least stay even with the competition? Or does the gotta-keep-up hype simply keep us all going, buying the latest gadgets or upgrades, jumping into the latest training or marketing opportunities on the Web? What does it mean to have our employability and even our identity hitched to the dynamo?

If we could control the structuring associated with the converging new information technologies, we could design and apply them towards an extension of all people and all human consciousness in all its diversity. Restructuring and the information highway could lead to a renaissance in community, cultural connectivity, and inclusive democratic participation. The problem is that the current direction and pace are for the most part out of our control. To the extent that they are being controlled by people at all, they are being driven by an elite of continental and global corporations and supportive government agencies. Restructuring could well fulfil Grant's lament both for Canada and for society as a whole: "As our liberal horizons fade in the winter of nihilism and as the dominating amongst us see themselves within no horizon except their own creating of the world, the pure will to technology more and more gives sole content to that creating."[52]

Grant's analysis is compelling, but by no means conclusive. He assumed technology to be deterministic and unstoppable. He also assumed that agency for change resided in the major institutions of church, state, and the corporate economy—not in women's groups, unions, local environmental networks, community organizations, and people in general.

To claim restructuring as a social and human issue requires first that we regain our own sense of time, as human experience, and escape the dynamo that is the official discourse on technology.

The Technological Discourse and Murphy's Law

The discourse as currently structured controls the would-be participants by imposing its frame on reality, its terms of reference, and its objectifying language. It mirrors the priorities of the global-economy dynamo by imposing that reality and those priorities as the context of discussion and debate, instead of the priorities of people and human communities. In this way the discourse achieves its most disinforming effect. It makes the human reality disappear.

The silencing of the human devastation associated with restructuring immobilizes people as people, reducing them instead to selective roles appointed by the discourse: the "expert" on "employment impacts" or "downsizing" or "structural adjustment," with everyone else consigned to the spectator stands. This stifles a truly critical discourse on technological restructuring, because it silences the voices of experience—not only all the voices that have tales to tell, but also the truths that they have to offer.

One truth is that technology doesn't just have an impact on people; people have an impact on technology. The relationship isn't entirely one-directional; it is reciprocal. People adjust the technology to make it work, to make it work better, or to make it suit them. The truth is that the technology isn't omnipotent, or infallible, although the technocrats and others who make a living selling technological fixes and cure-all promises would have us, and the discourse, assume so. In fact, technology is notoriously unreliable. About 75 per cent of large software systems either don't work the way they are supposed to or aren't used at all. Robots run amok. Systems crash. Files disappear. Software comes riddled with bugs. And viruses spread like wildfire.[53] These things happen, just as Murphy's law insists. (If things can go wrong, they will.) They are talked about over coffee and in office corridors, but they are seldom reported and rarely taken into account in the discourse on technology and its productivity promise.

Meanwhile, in what Ursula Franklin calls "the real world of technology," people cope. Using their often vast "tacit" knowledge, which combines personal job experience with that of co-workers, people find ways to get the system up again, to find ways around bugs, to fill in the gap between the often hyped-up promise of the technology and its often meagre actual performance. Workers' tacit knowledge has been obliquely recognized in the romance of National Secretary's Day and in the accompanying cliché about the whole office falling apart when *she* isn't there. That quality has only recently been taken seriously for what it is: a vital factor in making things and in making things happen; and a vital power in the hands and minds of workers. By grasping and building on this power, people can use it as leverage to gain greater control over the structure of their work and the technologies involved in it, and to resist being turned into servo-mechanisms of fully debugged computer systems. But instead this knowledge is just as likely to be tapped by management under new worker-involvement programs associated with "total quality management" and quality teams.

The most important truth that is covered up by the official discourse is that restructuring involves human beings in actual human communities—not two-tiered labour-market sets in virtual economies. It involves people whose identity, livelihood, and, at times, even health is being battered and destroyed.

Restructuring the world to serve the priorities of the global corporate economy, and in the process marginalizing people and human involvement into post-it-note adjuncts, or into permanent unemployment and workfare: this shift is morally wrong and unjustifiable by any human measure. That is the truth we need to resurface here. There is more to the real world than the global mutuals market, the IMF, and the corporate bottom line.

What is required is a new critical discourse, one whose language is the voice of experience, whose grammar is the rhythmic pulse of empathy and solidarity: a discourse informed by story-telling, with analysis and synthesis grounded in tacit knowledge and lived experience.

Such a discourse is already in evidence—for instance, in Laura Sky's documentary film *Lean Production*, sponsored by the Canadian Auto Workers, in Sophie Bissonette's documentary, *Quel Numéro/What Number?* and in *Voices from the Ward*, which documents the lived reality of restructuring through cutbacks in health care for people who work and are supposedly cared for in that industry.

This discourse is urgently needed, because the paradigm is shifting fast. By 1994 the investment in computers, communications, and other equipment by Canadian businesses had led to an increase of over 38 per cent in corporate assets since 1984.[54]

In mid-1995 Statistics Canada reported that nearly half of all workers (48 per cent) were working with or on computers, three times the figure of 1985. It also found that "the most elite class of workers, managers and professionals," were the most computer literate—with 75 per cent of the men and 61 per cent of the women in that group working with computer systems. The study also found that 14 per cent of employed Canadians were using the Internet, information highway, and other "high-technology lines of communication."[55] Some three hundred thousand Canadians are believed to be teleworking from their homes.

The new economy is structurally loaded and cleared for take-off down the information highway. Closure is approaching. It's time to check it out.

PART 2

BACKGROUND

Social and Cultural Transformations

3

Hype and the Highway: Virtual Corporations and the Agile Workforce

To say that any technology or extension of man creates a new environment is a much better way of saying that the medium is the message.
– *Marshall McLuhan*

A technology is not only a symbol of a social order. It embodies it.
– *Langdon Winner*

STUDDED in cover-story hype, the information highway entered the public eye as a giant consumer dream machine: a digital multilane expressway to the 500-channel universe and a teleshopping mall. In fact, the widest lane might well turn out to be the bus lane, filled not with people but with machine-readable data: sophisticated software programming robots at flexible multiproduct assembly plants, processing test results in labs, managing accounts and inventory systems in stores, banks and insurance companies, hospitals and liquor stores, collating student marks across school boards and provinces.

While the technological haves in the core of the new economy cruise the fast lane, schmoozing for contacts and contracts around the networked world, many others might be stuck in silicon pit stops along the way—call-centres and other digital worksites. Still others might be stuck at home as computer-controlled teleworkers or button-pushing couch potatoes, the 500-channel choice dissolving into a rerun of Ancient Rome's bread and circuses more than anything else.

The highway metaphor is mischievously misleading. Through its familiar iconography of roadways and on and off ramps, it would have us believe that what is being offered is simply one more line of transportation and communications running along outside the institutions of our lives. In fact, this highway is running right through them—through our homes, our places of work, our schools and hospitals. And it's transforming them through priorities and biases built into its internal structures and supporting software. Only a fraction of the controls are visible, like the knobs and dials on a dashboard. Most are out of sight and out of the reach of the end-user: in restrictive, arcane, or expensive access procedures, in flat-rate, pay-per, or package-deal pricing structures and systems incompatibilities, vertical and horizontal corporate integration, and so on.

The highway isn't simply a new technology, a new media, on the landscape. It is creating a whole new landscape and environment for living. And as McLuhan's phrase "the medium is the message" suggests, the medium of this new highway will fundamentally determine the meaning and message of our lives.

The Information Highway, IHAC, and Technological Restructuring

Whatever else its goal, the highway hype has helped to enlist public consent, if not support, for the massive public investments being earmarked to build a high-capacity multimedia communications infrastructure in an era of belt-tightening and government cutbacks.

The U.S. government has allocated up to $2 billion a year for its "National Information Infrastructure Program" and has linked "America's destiny," including its strategy for economic renewal, to the information highway build-up. In Canada, at least three provincial governments (British Columbia, Ontario, and New Brunswick) plus the federal government have also linked the information highway to economic renewal. The Chrétien government has allocated over $100 million to CANARIE, a private/public-sector consortium dominated by industry, particularly telecommunications and business-service companies, to build both the multimedia information infrastructure and the equivalent of gas stations and other support services to go along with it. It has also opened its $6 billion infrastructure-renewal program to info-highway projects. On top of this, the federal government has been spending over $3 billion a year on information technology since 1990.

Many of the government-sponsored information highway projects—such as a national research network and an educational network—are laudable as ideas. They will become so in fact, however, only if they are genuinely designed and operated on the terms of the people involved—as extensions of existing research and learning communities and cultures, not as commercial substitutes expressing the communications-as-commodity-model dominating the corporate agenda. And that's the issue. If the economic-renewal agenda around the information highway and information technology were truly inclusive and democratic, it could renew a richly participatory economy employing all the people who'd prefer not to be marginalized in, or dumped out of, the economy as we know it now. But if that were the case, you wouldn't have the federal government arbitrarily declaring forty-five thousand public servants "surplus" in 1995 while investing nearly $4 billion in information technology. And it wouldn't be turning counselling jobs in employment centres into computerized call-centre kiosks.

As it stands now, the information highway strongly resembles an economic-renewal project for the companies in the telecommunications and information-service sectors, which are getting the bulk of the government handouts, and a make-work project for consultants who are getting fat contracts. The corporate priorities are to find new markets for their information systems, to hook more customers into their infrastructures, and generally to fill the constantly expanding carrying and switching capacity of the capital-intensive information systems. In short, their priorities are in an important sense to colonize new sectors of the economy (especially health care and education in the public sector) and to recolonize others—in creating a new economy centred on the infrastructures of the information highway and dependent on its enabling software and switching systems. But these goals are not equivalent to the public interest. In fact, they are increasingly at odds with it, as the unprecedented divergence between rising corporate profits and declining wages and incomes dramatically illustrates.

"What's good for General Motors is Good for America" might have worked fairly well during the Fordist social contract of largely full and full-time employment coupled with decent wages and salaries for the working majority. But that contract has disintegrated with today's jobless economic growth, rising unemployment and underemployment, and the proliferation of marginal McJobs.

To the extent that it still exists, the debate about restructuring in this country is framed around the information highway. However, the hype has put the focus not on what's happening now, but on a set of promised entertainment, learning, and teleshopping channels and on a future that has yet to materialize. So there has been little scrutiny of the closest thing to a policy-making body in this field. The federal government's Information Highway Advisory Council (IHAC) was dominated by big business interests—telecommunications carriers, manufacturers, and big-business users such as the banks—and its recommendations represented a major retreat from the traditional Canadian mixed private-public approach to communications—an approach that viewed communications as culture and community-building, not simply as a vehicle for the transmission of information products. (I will return to this theme in Chapter 7.) Stressing the role of competition, IHAC's 1995 report proposed not only that business and business investment should control the highway construction, but also that "outdated and unnecessary regulatory barriers" should be removed to facilitate competition and business investment.[1] The business focus was so relentless that the one labour representative on the Council, Jean-Claude Parrot, wrote a minority report which was published as an appendix.

In keeping with a handing off of responsibility to market forces signalled in the 1993 Telecommunications Act, the government's regulatory role here would be largely reduced to ensuring the harmonization of technical interconnectivity standards nationally and internationally.[2] The government would ensure public "access" to the highway infrastructure only after business had built it. The model being pushed is evident in the report's core themes and key recommendations. One statement, that "fair and sustainable competition should be the driving force behind the Information Highway," speaks volumes.[3]

The report's first policy recommendation calls for the government to do whatever has to be done to "ensure the right environment for competitive development." Its second recommendation is that "the Highway network and new infrastructure should be left to the private sector, and the risks and rewards of the investment should accrue to the investors." The third calls for the highway to be "driven by existing or potential market demand." A fourth says that the highway should be technology-neutral; in other words, any media can be plugged in. And the fifth defines a "limited" government role as financial supporter, model-user, technical-standard facilitator, and intervenor to ensure "universal access" to the highway only when the market has not adequately fulfilled this role.[4]

As it stands, then, the highway will most likely emerge as an extension of the computerized production model that restructuring has been installing in the corporate economy since the 1980s—with that model applied to new areas such as education and health care and increasingly to the area of culture and so-called "cultural industries." This will fulfil the corporate economy's vision for the highway, too. For despite all the highway hype that they're doing it all for us as learners, as teleshoppers, as people wanting to expand our cultural and entertainment horizons, the information highway was primarily conceived for business and the corporate economy, not for culture and the community. Indeed, a conference on telecommuting, telework, and the information highway was originally scheduled as the first major federal government-sponsored conference dealing with this sphere of affairs.[5] As it turned out, that conference was delayed in favour of a higher profile conference held in February 1994. That meeting—which had more cabinet ministers, more media coverage—focused on the convergence of television and interactive games with computers and telecommunications, on the famed 500-channel universe, and the home ordering of pizzas and banking services.

To a certain extent, the Internet represents an alternative infrastructure model for the postindustrial society. A network of networks, including community freenets, public discussion groups, and organizational Web-sites and related home pages of multimedia information, the Internet has broken free of its original roots as a project of scientific researchers funded and linked together by the U.S. Defense Department. The Internet offers a user-driven infrastructure-development path more along the lines of community communications than commodity transmission. Programming and structuring power are distributed fairly democratically among end-users rather than concentrated in the hands of a few media conglomerates or information-service providers. But by the mid-1990s the Internet was also being absorbed into the sphere of influence associated with big communications and media corporations and the information highway model. Many banks and other commercial services were setting up shop on the Internet through Web-sites, while big companies such as Bell, Rogers Cable, AT&T, Microsoft, IBM, and Philips Electronics were providing Internet access and services.

A 1995 federal government publication on the Internet stresses the corporate economy: "Some private-sector businesses are justifying their use of the Internet on the basis that even if it's not a mainstream access and service-delivery mechanism today, the Internet or something like it will certainly be the way business will be done in the not-too-distant future."[6] Similarly, a government publication on the information highway described the technology first in terms of the corporate economy:

> The widespread use of communications and information technologies has already transformed banking and financial services. . . . The information highway will play a similar but even greater enabling and transforming role across the Canadian economy as a whole. . . . The key to competitiveness will be the ability of firms to develop, acquire and adapt new and state of the art information and communications technologies, products and services—the tools that will be available on and through the information highway system.[7]

The section on the public sector begins by stating: "Soaring costs in health care, education and training coupled with large federal and provincial deficits are stimulating interest in the electronic delivery of public services."

Harold Innis wrote of the "penetrative powers of the pricing system" to describe the commercial colonization of Canada by European mercantilists in the eighteenth and nineteenth centuries. Today we could talk about the penetrative powers of the information highway. Networked systems and services, serving as colonizing agents in postindustrial empire building, are being sold to hospitals, libraries, and schools as technical solutions to spending cuts. It's as though the right hand of government cutbacks *does* know what the left hand of business is doing.[8]

The history of the "highway" metaphor is revealing. U.S. vice-president Al Gore coined the image in 1993 when he announced his administration's massive $2 billion a year information infrastructure program to upgrade the existing communications infrastructure from low capacity (associated with copper and cable) to high capacity (associated with fibre-optics) capable of handling massive volumes of data, including multimedia. He was actually extending the imagery used in 1955 by his grandfather, who championed a national network of superhighways "to secure the country's defense and build its economy." The superhighways helped consolidate the shift from local and regional markets to national markets. This new highway project is helping consolidate the shift from regional and national economies to continental and global economies. The infrastructure bill was also a sequel to the High Performance Computing Act, which Gore championed as a U.S. senator in the 1980s to retool U.S. industry with "superfast computers and networks" that, in his words, would "enable this country to leapfrog the Japanese."[9]

There was no talk of consumers and entertainment associated with the information highway then. The whole thing was geared to giving U.S.-based business a great leap forward as a world power. It wasn't driven by consumers demanding a 500-channel universe with more choice of more-of-the-same commercial programming. It was driven by business and what *Business Week* called its "insatiable appetite for communications capacity."

This appetite marks the maturity of computerization and of the communications networking necessary to launch the corporate economy into a global cyberspace via the information highway.

The technologies that have garnered much of the publicity—cable and television integrated with computers for videos-on-demand, plus electronic publishing, interactive games, and teleshopping—represent both new industries in their own right and new markets for existing industries. But the central, enabling convergence is the one between computers and other communications technology. Computer power (in the form of microchips) provides the brains, while communications technology provides the connections globally and locally, bringing that brainpower to bear on everything—including the dispatch of movies and pizza orders, the treatment of patients in hospitals, and the manipulation of research and learning materials in libraries and schools.

Compared to the transformation of existing work, and the scope for even greater transformation inherent in the high-capacity information highway, interactive games and talking graphics are little more than the poppyseeds on the bagel, or the tail on the Trojan horse; although as gimmicks they have been usefully distracting.

The true nature of the decoy horse entering our social and economic space is apparent in some of the large corporate players emerging in the converging communications and media conglomerates. One of these is General Electric, which not only supplies the power to run all these high-powered systems but also, through its subsidiary General Electric Information Services Company, supplies a powerful set of operating and networking systems plus applied services, including Electronic-Documents Interexchange (EDI). In a move since paralleled in Westinghouse's purchase of ABC, General Electric is also positioned for the home-market aspects of the information highway through its ownership of the media giant NBC.

All along, too, business was quietly laying the foundations of the information highway within all the institutions of the old economy, through the computerization of work and the way it was organized in the 1980s. The move from a number of private and semi-private corporate networks through value-added network service providers towards one general-purpose infrastructure, which the information highway and the Internet have come to represent, is merely the last stage.

In the early 1980s, communications networks were being recognized as important, but only as the basis for new "network" industries, a subset of the information business. As a 1981 report, *The Emerging Network Marketplace*, concluded, "The concept of the network as a marketplace is clearly evidenced in the rapidly growing services being offered via many networks where medical, educational, manufacturing, processing, financial, agricultural and other [information] products are offered to customers."[10] By the 1990s, however, the perspective had shifted. Continental and global information networks were no longer just lucrative new industries. They were the context of a whole new economy, the site where the old industrial economy would launch itself into cyberspace as a digital postindustrial economy.[11]

This lift-off is the culmination of a great deal of technological change, much of it dating back to the invention of the microprocessor, which put computers into communications equipment and communication into computers. Computerized activities everywhere in the economy could then be integrated into sophisticated production, distribution, marketing, and other systems, all co-ordinated through communications networks for centralized management across larger and larger distances. It's a powerful combination: machine intelligence and intelligent communications networked together for centralized planning, co-ordination, control, and, simultaneously, a global distribution and scale of work. The restructuring has devastated employment and even conventional understandings of what employment is.

The network stage also involves a profound shift in social relations, in people's sense of community and even their personal consciousness. In this phase, when a wraparound corporate environment can be created anywhere once the necessary technology is plugged in, the headset attached, and the "enter" button is depressed, the machine is no longer outside us. It is all around us. We now live in the machine. When we look at the individual gadgets—computers, cellular phones—it looks as if the technology is still a set of machine tools in front of us like industrial-age technology. The difference is that these gadgets are only the tip of the technological iceberg. Most of the technology—the core infrastructure and operating software—is in the webwork of wires, fibre-optic cables, satellites, and built-in switching devices and operating codes running invisibly behind the walls and other skins of our lives and all around the world. Most of the control is being concentrated there too, including in the corporate structures and intercorporate alliances building the digital webwork and installing various levels of software to run it and govern its myriad applications.

The information highway is becoming the medium of this new political economy, and its "message" or meaning lies in its structures and the new wraparound digital environment they create. As Innis first stated the proposition, structures of communication—here, the information highway—structure consciousness. They do this as the biases built into them structure work, corporate organization, political, economic, and social relations, and culture. To an important degree they will influence and predetermine what we can and cannot do, what we can say, and what we desire.

Still, with the information highway only under construction rather than being fully completed, if we can critically grasp the restructuring that has gone on so far, there remains hope that we can change those structures, making cyberspace a vibrant and open "virtual" extension of existing human communities—a fulfilment of the promise suggested in community freenets and the Internet—and not a commercial and corporate regime running at close to the speed of light for no palpable human reason.

Restructuring and Social Impacts

The corporate restructuring that accompanied the integration phase of computerization established a consistent trend to consolidate tasks that had been brought into the orbit of computer systems and simplified by their software. A large portion of the new jobs created in this process were so structured as to be entirely defined by the computer. They could then be controlled by the computer too, which made it easier to monitor performance and to manage the work with a part-time "contingent" staff. Equally important, these choices also set the stage for work to be de-institutionalized further, to a global post-it-note workforce conscripted via the information highway. Not only can work be distributed to any places that feature plug-ins and technical accessibility, but it can also be controlled to an unprecedented degree of detail through remote management systems. Still, even in this otherwise deterministic story of restructuring for globalization, there remains, as always, an element of choice.

In offices the transformation ranges from the simple replacement of typewriters with word-processing computers to the integration of front-office automation and back-office automation (associated with data-processing, inventory, and accounting) into regionally, nationally, and even internationally networked management information systems coordinated from a few remote head offices.

- In one of the largest studies of its kind in Canada, involving two thousand school boards, libraries, hospitals, and municipal and provincial government offices, researchers documented a dramatic shift from full-time to part-time employment. "Permanent part-time positions had increased in 39 per cent of locals. Permanent full-time positions declined in 40 per cent of them." The changes weren't solely due to technology, but there was a high correlation: "Sixty-one per cent of the locals experiencing changes of 50 or more permanent full-time positions were locals with computer-related changes at represented worksites. Almost nine out of ten of those experiencing changes (escalations) in permanent part-time positions were locals with computer-related technological change."[12]

  Other changes included work intensification with no increase in pay and a rising tendency to transfer work out of union workers' hands as more and more administration was taken over by software. This work included payroll, auditing, stats, and related record-keeping, word-processing, and related preparation of documents; and bookkeeping, inventory, timekeeping, registration, purchasing, accounts payable, and other financial arrangements. At the same time, the study found that locals that had contracted out mundane computer tasks five and ten years previously—jobs like payroll and purchasing—were now doing this work in-house, though with a lot of the work carried out via software. What's more, it found, the software was increasingly generated not by professionals but by computer-programming pools, in which programmers worked within parameters set by computer-assisted engineering systems and using automated programming tools such as compilers.

- A study in a Toronto placement bureau reported a continuing trend towards part-time rather than full-time employment in the clerical field, with close to a third of the companies in the study hiring temps or part-time workers only.[13] The study also found a new emphasis on hiring "multifunction" workers, with higher educational qualifications but mostly with the ability to take the pressure of the "automated office."
- Another study noticed a polarization between computer-assisted and computer-controlled work along clearly gender lines.[14] The exclusively female data-entry staff spent almost the whole day at the computer terminal, with their work entirely defined, controlled, and monitored by computer systems, whereas the male-dominated data-processing staff worked with the computer systems and enjoyed much more freedom to move around the office.
- Similarly, a three-site B.C. study documented a general polarization of work between, on the one hand, professional-managers who used integrated information systems as tools of analysis and decision-making and, on the other, a generally deskilled and reduced clerical-support staff who worked increasingly for the computer system itself.[15] Some administrative work was being enriched through a trickling down of tasks from the professional and management ranks as these sections were simplified through computerization. But much of the work was being eroded and deskilled through computerization, especially at the integration stage. Although there were redundancies, the main effect was a pervasive computer-simplification of work, with no compensating opportunities for more enriched knowledge work in return. Furthermore, rule-of-thumb self-management and experience-based tacit knowledge were consistently displaced by credentialled expertise and expert-level software.

As one woman said, she "made an awful lot more decisions" before the computerization of accounting. "It was a lot more involved." Now the job is just feeding the machine: inputting information into the computer. It's steady typing, not type, pause, think, pull relevant cards, type, and so on. Another woman said that before computerization, "If we had to take a trial balance on our accounting machines that took about four hours of manual work. Now it just pops out of the computer—You press a button and it's there. And also . . . aging of accounts receivable is another function that probably took a whole day. Well, now it's just there, in the computer."

Like many others now working almost non-stop at the computer keyboard, the women complained of backache, muscle spasms, and headaches, even nausea from the relentless, routine work.

At IBM, office work-station systems, called PROFS, were installed on the desks of all professional staff.[16] What had been individualized secretarial support to the organization's managers and professionals was evolving into junior professional and technical support to the multifunction, market-oriented self-managing professionals. This is the new upper tier of the full-time clerical workers, and a university education is now a requirement. Tellingly, company policy dating back to 1979 requires anyone being hired full-time to have at least one university degree. As one manager said, "Today we would hire someone . . . with a teaching background or a degree in Business Administration or Computer Science . . . for a secretarial position because the job is not quote 'secretarial.' The job is far more systems oriented."

A program manager with the New Brunswick government's Information Highway Secretariat tells me that his secretary does "administration and project management, plus she sits on a conference-planning committee—things officers ordinarily do. She still sends faxes and things like that, things a secretary does." But, he says, her work has really evolved.[17]

It seems that what women gained in better pay and occupations through the 1970s and 1980s is being rolled back, leaving women working just as hard or harder, but for less. Pay equity and employment equity programs are being dismantled at a convenient time for business.

A law office case study adds evidence of the trend towards polarization in clerical, administrative-support work, with automation draining the creativity and personal autonomy out of many traditional tasks while mitigating this somewhat through a filtering down of new tasks from the professional ranks when expert systems and other software simplify work at that level.[18] It also signals how computerization is industrializing a lot of formerly intuitive, holistic professional work—in this case, legal research and analysis—and turning good jobs into little more than McJobs. Expert-level software assists legal research, case analysis, and document preparation. According to the study, "The software not only suggests alternative wording but also explains the consequences of making one choice rather than another. . . . These expert systems make it possible for junior lawyers and/or legal assistants to handle more complex problems."

The proliferation of expert-system support tools brings with it a subtle but important cultural transformation. The scope, and the time, for personal judgement is steadily reduced as reliance on preprogrammed algorithms advances. As Joseph Weizenbaum warned twenty years ago, computerized decision-making turns "judgement" into "calculation," setting the stage for simplistic prescriptions to eclipse the goals of human justice in the criminal justice system.[19] The proliferation of sentencing manuals in the United States, with sentences handed out according to a chart prescribing incarceration periods based on a variety of factors, suggests just how far this can go—and how much an overcrowded, underfunded judicial system can be enjoined to embrace it.

In the *insurance field*, what's happening is representative of computerization throughout the financial-information service sector. Software has taken over more of the transaction processing as well as related calculation and information-processing work, and it has allowed new self-serve features in which customers do the work for themselves. The restructuring is also coinciding with a general consolidation of financial-service industries, with a merging of banks, trust companies, stock brokerage, and insurance firms. The social impacts are enormous, and the most significant is probably the demise of these institutions as mass employers.

- A case study by the Université Laval on the insurance business in Quebec confirmed the delay of social impacts until the networked integration stage, which included the use of expert-system software to calculate the risk that the insurance was intended to underwrite.[20] As systems software automatically issued the claim notices, maintained the claim portfolios, and dispatched claim settlement statements, the proportion of clerical workers fell from 43 per cent to 26 per cent of the total in the four companies studied. Significantly, the number of insurance professionals rose only fractionally, although their share of employment rose from 40 to 58 per cent. A 1987 update revealed a continuing shrinkage in clerical ranks as well as a "hollowing-out" of middle management as integration made self-management by professionals and semi-professionals part of the new "entrepreneurial" or postbureaucratic corporate culture.

- A B.C. study found some disquieting trends in computer-monitoring and control.[21] Workers complained about the inappropriateness of monitoring as a measure of performance, because it ignored time spent away from the machine dealing with difficult and exceptional claims, but their complaints were ignored. "The place runs on statistics," one interviewee said.

  The study noted that clerical workers were denied access to many data bases that would give them the information they would require to take their jobs beyond rote data placement. Access to data bases is subject to approval from the Systems and Procedures Department.
- An interview-based study in an OHIP claims-processing office with an all female data-entry staff found that computerization might have first enriched the work to be done, but as the bugs were ironed out of the system the jobs often ended up being worse than before.[22] While tasks had previously been varied, with some scope for individual programming in the early stages of computerization, later they became more standardized. The women reported that their jobs were more tied to the computer. One claims clerk said, "Everything is programmed. You have to do what the machine wants. The other one, it was more free. First of all, you made the program yourself."

  Furthermore, once the work was computerized, "The keyboard becomes the actual site of work measurement." This not only permits full computer-monitoring. It creates an airtight job ghetto. As one woman explained: "I used to go really fast because I thought they'd notice me and maybe I'd get promoted. Then one day it sort of hit me—why should they move me? I'm a top performer. It's sort of a Catch 22."

Monitoring is also used to motivate and discipline workers' performance, with hidden discriminatory effects. In June 1985, management set a minimum performance level of 11,400 key strokes per hour, warning that anyone falling short would receive counselling and training, and anyone still falling short after three months would be demoted and possibly "released" from employment.

The women said they were "insulted," "humiliated," and "degraded" by the memo. They also decried how the focus on flying fingers to the exclusion of everything else created an age-related bias by which only the youngest and fittest could compete. As one woman said, "You slow down with age."

Finally, the report chronicled many of the stress-related complaints that have come to haunt people confined to concentrated keyboarding, with little or no personal control over the work situation. The majority reported chronic back and shoulder pain, eye strain, and problems with wrists and fingers; and, for a sizeable minority, nervousness and depression. Significantly, one segment of the study looked at the male-dominated occupation of data-processing technician. It noted that although the workers there spent almost as much time on computer terminals as the women in data entry, the men suffered far fewer of the stress symptoms than the women. But they also controlled their work situations much more and were free to move around the office, while the women were not.

In hospitals computerization began with ad hoc computer installations and software add-ons. One hospital, for instance, installed an electronic inventory system that it used to generate equipment-purchase forms, a room-by-room inventory checklist, forms for instrument quotation evaluations, plus vendor lists and specifications. These were all derived from the same data files, with little human involvement.[23]

In the next step, hospitals integrated patient registration with other modules of background information and administration, then added other steps along a patient's path through the hospital system. In hospital labs, "blood work" shifted from performing tests to merely recording test results. Some software can both execute the test and analyse the test results, making a preliminary diagnosis, which will be relayed automatically through the computerized information system to other specialists. In intensive care, computerized systems monitor vital signs (pulse, respiration, blood pressure) and spew out trend analyses based on their "expert" level of software.

Step by step, the systems have converted the information base of caring for people in hospitals into a stream of electronic data. Food-allergy information recorded at the registration terminal can be automatically incorporated into food-tray assembly terminals in the kitchen, and outside data bases and experts can be tapped through electronic network services offered by the phone company. The system has also enabled a complete restructuring and reorganization of hospital work, transforming much of it into computer-defined and computer-measurable jobs that can now be done by a less qualified, part-time staff.

The patient-care management system has completed the circle of this transformation. It unites all the computer's background information and management capability around the actual handling of patients in hospital wards, orchestrating the interaction between nurse and patient according to directives coming up on its computer screens. The system encloses nurses in a computerized redefinition of "patient care" based on a set of time-managed job tasks as listed in an electronic patient-care file.

The system is flexible enough that if a nurse wishes to do something that hasn't been specifically ordered by the attending physician, she (or he) can key it into the system. As long as it has been preauthorized as among the procedures not requiring formal authorization, the system will post it as a new "patient-care order" in the "patient-care schedule." In the new hospital-management emphasis on "value-added" action and quality control, there is a telling bias in what gets listed as a legitimate patient-care event and what does not. Giving a bath or a back-rub just to comfort and console a patient—taking care as opposed to performing a potentially billable treatment—tends to be excluded. This is a bias not only towards prescriptive action over the more holistic action of promoting health and well-being, but also towards opening the way for the more commercialized, service-product approach to medicine that prevails in the United States, where health care has always been an industry.

The system's software supports a number of administrative functions, such as fine-tuning staff loads and shift schedules. By its very nature too, the system provides an electronic trail of all the procedures, including all the people, materials, and equipment involved in patient care. It was designed for billing various "order charges" and also as protection against costly malpractice suits in the American context of these largely U.S.-designed and managed systems.[24] In Canada its use could easily shift from cost-accounting to tagging various "value-added" services provided, possibly, by private clinics or expert consultants available through the information highway networks but not covered by medicare—as well as for flagging what features the government is willing or unwilling to count as a public health-care service.

The most dramatic effect of computerization and restructuring in Canadian hospitals has been a shift from full-time to part-time staff and a stretching of full-time staff to their limits.

- A survey of twenty-two hospitals in Ontario concluded, "The most unambiguous and disturbing finding which emerges is the widespread decrease in full-time positions" and a corresponding reliance on part-time nursing assistants, nursing aides, and non-specialized technicians and maintenance personnel.[25]

- A Quebec study of clinical laboratories found a dramatic shift from full-time to part-time employment, an overall declining level of employment corresponding with increased workloads, and increased hours worked per week. Those increased work hours were explained by the growing use of temporary workers and the increasing permanence of an "availability list" for temporary work. "We conclude that permanent full-time positions are tending to disappear in favour of permanent part-time positions and above all, temporary jobs."[26]

  The study didn't draw a direct causal link between this shift and the increasingly computerized equipment used for laboratory testing, which occurred at the same time. But the coincidence of the two developments supports the pattern documented elsewhere of an increasing specialization of work tasks along an industrial hierarchy, with the technology tending to polarize work into a deskilled technical support staff who, here, support and service the automated testing equipment, while a highly skilled professional staff interprets and analyses the results.
- An interview-based study by the National Federation of Nurses found that nurses had a sense of being torn between the computer and the open ward as the focal point of their work.[27] There was also evidence of a struggle for control over the definition of the work to be done as nurses sensed this being drained away from them into the seemingly neutral frame of the computer screen. The executive summary noted that only "a very small percentage of nurses report improvements in the delivery of nursing care as a result of computerization . . . perhaps [because] most of the computer systems used by these nurses do not support the delivery of nursing care in ways that are valued by nurses."

  One nurse commented, "I often feel that the computer takes priority over the patient."

  The report noted that nurses sometimes ignore the clinical test results routed to them via the computer system, choosing instead to rely on what they considered "more up to date [personal] observations of patients."

By the early 1990s, however, nurses were losing that option—of spurning the systems-informed approach to patient care in favour of their own traditional knowledge and direct personal relations. They were being squeezed by a new management program called "total quality management," which inculcates a systems methodology and mindset as the new culture of the workplace.

Total quality management (TQM) is a Japanese management philosophy stressing process innovation over product innovation—a reflection of Japanese industry's postwar tradition of "reverse engineering," in which the Japanese fine-tuned technologies invented by others. It focuses on "continuous improvement" of the production process, striving constantly to upgrade the "quality" of that process. It defines quality as "performance to requirements."[28]

While the rhetoric of quality suggests the art of good patient and customer relations, its substance is the new science of computer "metrics." One motto of this science is "If you can't measure it, you can't manage it."[29] This implies that if you can't count it, it doesn't count.[30] If it doesn't compute digitally, or commercially as a billable intervention, it doesn't compute as real.

This metrics-centred management philosophy is now central to the work-reorganization phase of computerization. It is both a feature of the new systems-driven workplace and a program for adjusting people into compliant, functioning parts of it. Nurses and nursing aides now have little choice in the matter. The systems definition of work and its forms for recording the results are the new context for measuring worth on the job, although this can have some sanity-warping results, as a study called *Voices from the Ward* makes poignantly clear.[31]

"Every job is timed. We have a time limit on everything," a dietary kitchen aide says. In the operating room, for instance, three minutes are allocated for clean-up between operations. A physiotherapy assistant's day is divided into fifteen-minute modules, and she has to account for each one of them in daily reports.

The study dramatizes the contradictions between affective rhetoric and effective reality: "Quality Control is about filling in sheets about standards you cannot meet."

"Patients [are] not having their hair combed . . . or having somebody take them to a bath. There is no dignity, no privacy, there is no nothing. More often there are bedsores, rashes and even people falling out of bed."

There are no limits on surgeries, which are automatically considered "value-added," even, the report notes, when they are unnecessary. Yet such strict limits are set on the availability of diapers for the geriatric wards that patients have frequently been left in wet diapers for hours.

Driven by the dynamo of hospital-management systems, health-care workers are being driven to abandon their standards and becoming stressed out to the point of personal breakdown.

One nurse asks, "What happened to our judgement along the way? That's what I feel. . . . They're picking us to death."

It is all taking a toll: "Last week . . . every night I came home, I had a couple of drinks. . . . It's gotten to the point, I'm going to explode."

Another: "It's dog eat dog. Everybody is nervous. Everybody is tense."

Another: "I'm a big woman, but it's either talk back and get fired, or cry."

Computerization, notably in the choices made around it, is also taking its toll on the culture of health care itself, making it more of an industry. As one nurse says: "We're not a hospital anymore. We're [supposed to be] there to look after patients. I find they're being lost in the shuffle. We're being lost in the shuffle. And it's become a business to the point where all it is is money."

The more work is standardized, with prescriptive procedures replacing holistic involvement everywhere from hospital wards to specialized clinics such as physiotherapy, the easier it is to implement the new work model of a two-tiered core/contingent labour force. In this scheme, less skilled aides and technicians are hired part-time to work under the overall supervision of the hospital's patient-care and management information systems, augmented by highly credentialled directors of nursing and specialty areas. Furthermore, as these systems are standardized across whole continents, it becomes that much easier for "virtual corporations" working the information highway to plug in new management and consultative services, extending a process of privatization and industrialization of Canadian health care from within.

The fact that the bulk of these new computerized systems and services originate in the United States is merely another detail of the transformation. It then becomes a small incremental step for U.S. health-management systems and health-insurance companies to start running as well as managing specialized clinics—for instance, in physiotherapy—and expanding these services into "virtual surgery" and diagnostic consultations over x-rays and other data-defined test results, all through information highway hookups.

In *retail services* what's happening also has implications beyond its own industry, stretching out to what's happening in personal and consumer services. As such, the transformations have implications not only for workers and working but also for the culture of human communities.

Two computer-linked technologies have revolutionized retail service, blurring the distinctions between production and retailing and creating a whole new time-centred standard of competitiveness. The technologies are the bar code on products and machine-readable scanners at computerized checkout counters, which are linked on-line to head-office computers. Computerization swept through retail checkout counters through the early and mid-1980s, bringing with it what *Business Week* magazine called an "automated management system." Value-added software analysed the data pouring in through the scanner and out through the cashier's keyboard, allowing stores to trim inventories through a "just-in-time" system of stock reordering. It also allowed them to rely on less qualified part-time staff in many areas and to trim staff to an almost just-in-time scheduling basis.

- A B.C. study noted that computerization had permitted a massive shift from full-time to contingent employment, resulting in profound cultural changes.[32] What formerly had been fairly stable relationships among staff, with a familiar in-store community, had become fractions of relationships among part-time employees, meshed together more by the computer-controlled assignment of tasks and time-slots than by collaborative decision-making. As the study reported: "Part of what it means to have a job is . . . a stable relationship with a group of work mates. Most respondents reported that increasingly erratic shift scheduling was a problem in this regard." The study found that 53 per cent of clerks and 78 per cent of cashiers now worked part-time.

Even when the employees were together, they had little to say to each other. An assistant manager noted: "I don't have to think about anything anymore. You know, it's moronic." The weekly staff meetings were being cancelled as redundant.

The computer is not just looking after stock and shelf management. It is also supervising staff. Of the over three thousand Canadian and U.S. companies reporting to the Food Marketing Institute in 1984, fully 90.3 per cent listed cashier-monitoring first among "the current uses of scanning [point-of-sale] data." These data are then used to "manage" employees, with high-scoring employees given more work hours, and below average scanners subjected to disciplinary action. And customer relations are subtly transformed. Just as the statistics exclude the time spent with a regular customer who returns to a familiar cashier for a friendly chat and perhaps some help and explanation, so those aspects of being a cashier that aren't computable—providing "customer service" by interpreting the store and its complexities to customers—are marginalized and forgotten.

Computerization through the 1980s turned stores and often their suppliers into programmable modules in corporate management systems. The Italian clothing chain Benetton then went the next step, launching those modules into cyberspace through real-time electronic networking. It created a continuous feedback loop, channelling sales information from its retail outlets directly into its production-planning department, which in turn contracted out up-to-the minute sweater orders, detailed by whatever colour, style, and size were selling fastest in the stores. Home-based knitters made up orders precisely according to these trends. Initially, Benetton provided the communications system. Since then the company has turned over the networking part of the work to General Electric's Information Services Company.

By 1991 Benetton had tripled sales in seven years. It had also set a new time-centred standard of production and market delivery called "quick response," which has collapsed production cycles from weeks to hours and days. Companies like Du Pont on the fabric side and retailers like JCPenney and Wal-Mart have emulated this model. By 1991 Wal-Mart had invested half a billion dollars in "quick response" point-of-sale equipment with connections to the information highway.[33] Companies without this networking capacity were struggling or going out of business.

Quick response is more than a fast-paced way of making and selling things. Through the networks and networking on which it depends, the system represents a new way of organizing production and getting it done: a model of integrated production, distribution, marketing, and even consumption based on continuous marketing and market feedback. The technique of quick response is transforming mass production into demassified modules of "customized" mass production, featuring "virtual products" that customers help make for themselves by choosing from among multiple-choice options available on teleshopping lines and channels and websites. These customized orders can then be beamed along the information highway to computerized workshops or factories on the far side of the world. There the computer simulation will be transformed into a material product, while the reality of potential forced labour and dangerous working conditions is conveniently kept out of the picture.

Virtual Corporations and the Virtual New Economy

The networked information systems of the information highway blur borders between countries, and also between industries, industrial sectors, and even between goods and services. A display at a trade fair held in Toronto in 1994, for example, featured a remote-control mining operation jointly sponsored by Inco and Bell Canada. The company signs on the display were about equal in size, which seemed appropriate. In the demonstration a single operator used a remote-control machine to activate and direct two scoop trams in a nickel mine 450 kilometres away and 900 metres underground. What had been a heavy and hard piece of mining work had become information processing handled by a technician who might never have to visit Sudbury and who could be employed by either Bell or Inco.

In transportation, American Airlines developed an airline reservation-handling system, called Sabreline, and marketed it as a separate information service. By 1990, 93 per cent of all travel agents in the United States were using computerized reservation systems, and a third of these used this system. The system makes money for the airline and now represents 25 per cent of its corporate value. The information collected through Sabreline also gives the company economic leverage—for instance, in accurately predicting what flights will be full and what flights require more discount seats to achieve capacity.[34] In other words, information becomes "both a commodity and a form of control central to the creation of new commodities."[35]

Information, as market feedback, is blurring the distinction between production and retailing, and revolutionizing both in the process. This is why the telecommunications companies and the computer systems, software, and network service suppliers that are all clamouring to build, run, and service the information highway are becoming so important in the new economy. The software in the enabling systems controls how work will be defined and where work will be done, and the management of its feedback determines both what new economic activity and what new employment will be created.

As goods become information in merely a shell of materiality, the difference between goods and services disappears. Service is becoming a larger part of goods, and goods production is being integrated into an ongoing service relationship. Interestingly, the Harley Davidson motorcycle company is seen as a pioneer in this new model of upgradeable products, because modular add-ons have traditionally been part of its philosophy of continuing customer-service relationships that help cement customer loyalty into near dependency.

As this happens, too, the difference between paid work as employees and unpaid work as consumers is blurring. More work is being privatized into the hands of customers who take on the role of unpaid assembly workers. The phenomenon, in turn, exacerbates the competition for scarce jobs while intensifying home life and leisure.

But the biggest changes are in the organization and corporate control of work. The new economy not only de-institutionalizes work and economic activity but also de-institutionalizes corporations. Globally networked electronic infrastructures offer them a new virtual corporate skin, re-institutionalizing them around the information highway. In essence, the highway reconstitutes them as virtual corporations and even as virtual cartels called "federated . . . virtual enterprises."[36] According to promoters of these monopoly-scale virtual corporate agents, these enterprises are "likely to become the dominant world industrial order of the 21st. Century."[37] Through the information highway and its variants, such as the Factory America network (discussed in the next chapter), these companies can extend their management information systems, and, through them, their corporate culture and controls, into plugged-in sites around the world.

Throughout history, communications networks—especially fast, space-binding ones—have expanded spheres of market and other influence while consolidating centralized control. From the days of merchants using networks to break the bonds of municipal markets in the Middle Ages, networks have regularly changed the "scale of affairs and broken its limits."[38] So what's happening is nothing new. What's new is the level of scaling-up in time and space.

As one commentator put it, "The real benefits of common channel signalling techniques, packet switching and ATM (asynchronous transmission) technologies come from their ability to integrate and exploit whole networks, dynamically responding to demands and using the network as a unified system."[39]

The virtual corporation is the creature of that dynamism and of the new "business paradigm" defined by "high-speed capital, technology transfer, access to cheap labour and collapsing product and life cycles."[40] Supported by the makings of a new, largely "invisible" world government associated with organizations like the International Monetary Fund, committed primarily to ensuring stable financial markets,[41] these new corporations are moving beyond local, national, and certainly democratic control.

Globally networked corporations have broken the bonds of the nation-state—certainly, of smaller nation-states like Canada, Mexico, and most of the South American countries. Through restructuring, deregulation, and free trade, the corporations have broken free of formerly restrictive social contracts. They have become "a political and economic force" of their own. The idea of a "federation" of such electronically networked virtual corporations and enterprises housed in and outfitted by the information highway is suggestive of this new political and economic force. It's one of many ideas emanating from the Agility Forum, a U.S. think tank on continental industrial renewal. Founded at the end of the 1980s, the Forum is an industry-academic collaboration with financial support from the U.S. Defense Department. It has brought together senior executives from the major military industrial suppliers, plus the major companies in computers, communications, and the auto industry, some with branches or subsidiaries in Canada. It has sponsored research and industry brainstorming conferences and published reports extolling its vision for a digitally reindustrialized North American economy. This postindustrial economy features agile workers in flexible production facilities, managed and co-ordinated through the invisible structures of the virtual corporation and the "federated enterprise."[42]

The documents don't talk about monopoly capitalism having reached its omega point through the virtualizing infrastructures of the highway networks, in which monopoly corporations and conglomerates vanish as structures altogether. Instead they mildly predict that virtual joint ventures will become "routine,"[43] with one dissolving today and a different one formed tomorrow and many happening simultaneously, linking different parts of different corporations in time-compressed, concurrent production projects. The alliances—called "the new intimacy" between "erstwhile competitors"[44]—are made possible by "streamlined legal practices," "standardized interfaces," and a "pervasive connectivity"[45] through the increasingly standardized networking systems that the forum is promoting. Integration is "electronic," as companies agreeing to act as suppliers or consultants or distributors on this or that product or "service product" are given access to certain data files, software, and other technologies to allow close project-centred co-operation.

One document, *Key Need Areas for Integrating the Agile Virtual Enterprise*, offers a game plan for streamlining everything from technology and legal protocols to people in a "contingent workforce," so that these production and human modules can be plugged into any collaborative "multiventuring" corporate combination for work along the information highway at no start-up cost to the company involved. The game plan also calls for more and better computer "metrics," so each corporate module's "value-added" contribution or each individual's hourly performance can be measured for "value-based compensation" and profit-sharing.[46]

The report envisages a "certification process for organizations" to qualify to serve as "suppliers" in "cross-functional teams" associated with virtual business arrangements—the quasi-middle class of supplier companies in different countries and regions—but with an implied price of admission in credentials, hours, pay, and other standards being controlled by the larger virtual enterprises. One part of that price seems to be a commitment to ongoing "business process re-engineering"—that is, the perpetual effort to speed up the flow of everything from handling orders to making products, plus getting products to customers and compiling feedback for the next generation of service products. The process would drive everyone faster and faster through the production-consumption cycle—*and all supposedly for us*: the fleshy feedback function in these global technological dynamos.

The report also suggests a "resource inventory of skills and competencies." A stepped-up and global electronic version of a "suppliers" list and possibly combining elements of a temporary employment agency, this inventory would take the form of a centrally co-ordinated data base containing "human and information resources, tools and techniques, manufacturing and other assets central to a (virtual) enterprise." This list would be updated through a recertification process and accessible to participating companies.

In matters of law, the report insists, "The anti-trust laws must be relaxed to facilitate the move toward cooperation, including data sharing, protecting proprietary data, trade-agreements assessment, enterprise zone definitions and foreign tax credits.

"Similarly," it says, "product liability laws and practices need to be referenced to better balance reasonable customer protection with extended risks for the enterprise. Participating organizations in the enterprise must share product liability risk."[47] This suggests a move to diffuse corporate accountability across the entire webwork of participants, so the buck virtually stops nowhere.

The biggest shift away from social and local accountability is probably the one affecting working people. The Agility Forum's key-needs document clearly envisages that "the role of a temporary or contingent work force will increase." Furthermore, "Laws considering minimum wage, equal employment opportunity goals, and organized labour's role need to move toward less controls and greater recognition of value-based compensation."[48]

By 1993 nearly half of Canadian companies offered some variable pay based on the performance of the individual, the work group, or the company. Unions worried that this would promote speed-ups and competition among workers.[49] By that time too, a two-tiered employment scene, buttressed by a third tier of the frequently and long-term unemployed, had become a new feature of the social landscape. A rising proportion of women and of both men and women under the age of thirty-five had been entrenched as the new "contingent" segment of part-time, contract, term, and temporary workers, with no job security, no continuity, and few if any benefits, let alone union protection. Furthermore, both developments had been so relentlessly portrayed as simply new aspects of a labour market associated with a more globalized and services-centred economy that the public had accepted the changes as normal and socially acceptable.

At the same time technological restructuring had reached the point at which more and more work could be defined and controlled by information systems and networks, so more and more people could be employed only peripherally. Now, as corporate networks branch out through the information highway, more and more workers can be employed in de-institutionalized and even privatized settings, employed "virtually" in any place that can be turned into a virtual workplace through the simple device of a phone and modem. Corporate management information systems can do the rest.

It is important to distinguish here between telework and telecommuting. *Telework* is usually reserved to describe people who are employed at home doing generally routine information-processing work, often under poor pay and working conditions, and with little power to do anything about it. *Telecommuting* is associated with people who have a regular office and occasionally work at home on a lap top or personal computer, usually under conditions of their own choosing. The 1991 census recorded a 40 per cent increase since 1981 in the number of people working out of their homes (compared to a 16 per cent increase in the labour force as a whole). Most of these people were in the information sector: management and administration, clerical, sales, and service work. The pressure is clearly on, with companies like Rogers Cable and Northern Telecom selling systems, such as "CableLink-Work," and software, such as "Perfect Office," along with their other offerings, enticing people to ramp up and log on from home.

Depending on the trends, that portion of the new work called telework—that is, work downloaded by computer networks and completely defined and controlled by management information systems—could become the new post-Fordist model of work, a pillar of a new cybernetics of labour.

The current structures permit immense decentralization of operations and connectivity, but they also encourage immense centralization of control and, with that, a new standard of operational performance. The global networks and monopoly-scale companies offering infrastructure, networking, and other information services promise new economies—not just of scale but also of speed and scope. The scope includes concurrent operations (design, production, marketing, market research) and the package-deal integration of everything from systems software to networking plus on-line services and information products. The companies involved include computer giants such as IBM and Microsoft, communications/utilities giants such as AT&T, Bell, Stentor, and Rogers, and media conglomerates such as Time Warner and NBC (General Electric).

Given this unprecedented new meaning of scale, the information highway could herald a new wave of colonization, with the highway as the medium of colonization and its globally networked systems and service providers as the agents. These systems will colonize through the new standards of fast-forward, quick-response, agile, and concurrent production. They are also colonizing through a new set of economic relations, such as virtual alliances and contingent, post-it-note forms of hiring. They're doing it through the global distribution of workloads across a vast network of call-centres, virtual factories, and teleworkers located far and wide in their virtual workplaces. And they're doing it through a new language—the enabling language of software and a networked, computerized way of doing practically everything.

Equally, and in keeping with the meaning of colonization, the global information systems are displacing the old economic relations. They are replacing old institutions such as fixed bricks-and-mortar factories, hospitals, clinics, and offices, and old infrastructures of transportation, communications, and distribution. They are attacking old standards associated with local and sequential research and development, with holistic human services and traditional industrial and human relations. They are discarding the old language of engaged human knowledge, skill, innovation, and empathy.

If current trends continue, hospitals and clinics, schools and school boards, colleges and universities, and public and university libraries could be colonized too, through their increasing dependence on computer systems and services. Like the subtlest forms of colonization, this takeover won't happen at the dramatic level of imposed ideology, but at the mundane and almost technical level of daily survival. Budget cuts will be bridged through the adoption of computerization and remote information, management, and other services, with downsizing measures included. The process will be colonization, and simultaneously privatization, from within. As local activities and knowledge are hollowed out in favour of remote data bases and research services via the information highway—by libraries, for instance—employment in local institutions will be hollowed out too.

The colonization from within will most likely mean a continuing hollowing-out of employment across all public services. Software will be constantly upgraded and expert systems refined, and a number of different suppliers and distributors will be replaced by a few integrated service suppliers, bypassing local distributors, consultants, and other intermediaries.

The marginalization of people and the downward drag of earnings and benefits will continue in lockstep with jobless economic growth. Meanwhile, the new working rich and what's left of the middle class will be kept busy at the core of the new economy, putting in ten-hour and twelve-hour days—or up to twenty hours in the four-month "product-development" season of the high-tech sector—feverishly engineering the next generation of products, marketing those products, and finding clever new niches for them in the ever-expanding empire of cyberspace.

For the transnational companies plying the information highway, free trade and deregulation have removed obstacles—such as the need for local investment and local employment, and employment-equity policies—that traditionally held them accountable to local communities and countries. With these regulations removed, they are free to operate globally at full capacity, distributing workloads from one work station to another or one computer system to another as costlessly as possible. They sought and have largely achieved a level playing field across whole continents and hemispheres, pushing for the same (generally low) standards regarding labour laws, unemployment insurance, and other "payroll taxes," as they call them. Globalized production also requires a global levelling-out of knowledge and operating procedures, and the displacement of quirky local ways and means of doing things that don't fit the new standard practices. Free trade combined with government cutbacks, deregulation, and privatization have achieved much of this preparatory work, permitting global-scale corporations and conglomerates to boldly go throughout the cyberspace of the new economy, wherever a new market or investment opportunity beckons.

4

Across the Digital Divide: Manufacturing as Global Agility

No class in history has ever risen faster than the blue-collar worker, and no class has ever fallen faster.
– Peter Drucker

Progress is accelerating.
– IBM Research Centre executive

Work is timed right down to the second. It's continuous work. It's stressful.
– GM worker.

IN AN AUTO-INDUSTRY MAGAZINE of October 1993, a sketch accompanying an article on agile manufacturing catches both the promise and some of the hidden contradictions. The picture features three trim, well-dressed young men juggling geometric objects while riding unicycles on a circular platform poised somewhere above the earth. One of the men, the focal point of the picture, has his tie flipped over his shoulder, he's riding so fast and free. And that seems to be the message: these are modern-day cowboys riding the global cybernetic range, taking on an ever-changing series of challenges around the world, and riding them!

The catch is that these free-wheeling athletes are confined to the face of a clock. You don't notice it at first, but the platform they're cycling on is a clock. The clock face contains them, circumscribing their mobility. A large minute hand sweeps them along at the pace of an undisclosed control mechanism hidden beneath their peddling feet. For me, the message is that Taylorism isn't over; it has merely changed perspective, from external to internalized rules as subtle, seductive, and potentially debilitating as the "outperform" hype of mutual-fund ads.[1]

The use of quick response (QR) in retailing and the military set the stage for agile manufacturing. It began as a marketing strategy by which companies like Benetton began relaying real-time buying trends directly back to supplier-manufacturers using digital information scanned off bar codes attached to every item on sale. It became an industry standard as the idea was taken up by larger corporations, for example, in a collaboration between a U.S. textile consortium, TC^2, and the clothing manufacturer Millikans Industries. These companies together developed a QR software package for automatically relaying point-of-sale data from in-store computers back to garment and even fabric producers. QR collapsed production cycles from weeks and months to days, which increased sales. One report noted that "retail sales were flat" in 1990 except for a hundred large retailers and manufacturers using QR. Their sales were up by 15 per cent.[2] The companies included JCPenney, Wal-Mart, and, in Canada, the Eaton's chain, which is linked to the clothing supplier Grand National Apparel. Not only does information move back and forth instantaneously, allowing Eaton's to implement an "automatic replenishment" policy for its Haggar apparel lines, but employees of both firms are also involved in "cross-functional/cross-enterprise teams"—though probably not with the same numbers of employees in each firm. Such teams are also an example of the new virtual corporate enterprise, with "integration by bytes instead of bricks."[3]

Quick response is also creating what is called "virtual products" as it transforms what had been static material products into a fluid blending of product plus service. One feature of this is customizable and upgradeable goods or service "products," in a "cradle-to-grave" product-support service system involving insertable cartridges, software updates, and other "add-ons." Promoters advocate "hooking" customers into "long-term relationships" and people "buying into a relationship" with a supplier, which "binds consumers to manufacturers."[4] In practice companies are also offloading more and more work from paid staff to consumers. Alvin Toffler's term "prosumer" applies to this blending of producer and consumer roles, and it amounts to a wholesale privatization of work. Consumers become participants in the production process as they install the next generation of software or work their way through computer-linked voice clips at the end of a 1-800 customer-service line. They're contributing to jobless economic growth as they help produce the next model—of their computer, VCR, telephone, or car—by reading the instructions, following the software package, installing the upgrade themselves, and providing feedback on bugs and other shortcomings to inform the next-generation product.

The postindustrial revitalization of industrialism associated with corporate and global restructuring is moving towards this pattern: rapidly upgradeable, customizable variations on standardized parts added onto core generic subsystems. "The end result is not a truly custom-made product, but a customized configuration of standard components," says Steven Goldman.[5] Equally important, these components are manufactured in instantly reconfigurable production processes to serve a more fluid, information-centred conception of production structured as much around consumer relations as conventional production relations.

The approach also incorporates strategic military interests, at least in the United States. QR became a new operational norm in the military-industrial complex during the Persian Gulf War. One example was in the development of a "friendly forces identification beacon," which yielded a prototype test unit within five working days through extensive subcontracting among commercial suppliers.[6] In another example, Lockheed and Texas Instruments took four days to develop a so-called "smart" superbomb capable of destroying the seemingly impregnable Iraqi command bunkers, and they delivered the final product in thirty-seven days.[7]

Texas Instruments became one of several military-industrial representatives on a sixteen-member "Inner Core Team" of high-tech and automotive industry representatives (plus one Department of Defense representative) associated with an initiative funded by the Defense Department to develop an "agile manufacturing" agenda for U.S. industry in the early 1990s. The core team's two-volume report speaks of "an agile manufacturing system developed in coordination with DOD needs." It spells this out by defining agile production facilities as "designed to require no special modification to meet DOD requirements."[8] In other words, the reconfigurability requirements will hook the core of the U.S. "domestic" economy into a state of permanent war-readiness: reconfigurability with an underlying military or paramilitary preparedness design feature so that production facilities of any kind can be reconfigured to produce weapons for the military—or, presumably, for the police and prison system as the war on drugs and crime heats up demands for bigger and better instruments for crime control and public order.

This is a frightening application of the so-called "peace dividend," in which "turning swords into ploughshares" means burying a sword in every ploughshare. It also means that a military production model with its built-in bias towards remote control in a presumed hostile operating environment is being diffused throughout the "at-peace" economy.

This scenario is the project of only one, albeit powerful, interest group in the U.S. economy. The documents simply represent its agenda, and that vision can be resisted in a host of ways. But to be resisted, it must first be understood in light of the production model associated with agile manufacturing.

The Stepping Stones to Quick-Response Agility

The Third Wave shift from machine-based mass production to computer-based, information-intensive agile production involves a continuing process of standardization and, at the same time, the ongoing transfer of intelligence away from human hands and minds into computer systems. This is also a dialectical process, full of contradictions in the daily practice of technological change. Still, this process has moved steadily along, through a number of identifiable stages.

Computer-integrated manufacturing begins when individual applications of computerized technology—robots for materials handling and assembly, computers for testing and production-scheduling, automated materials-handling systems—have reached the point at which they can be plugged together to form integrated production systems, with sensors and machine-readable forms providing the connective tissue. In Canada the largest companies were at this stage in the second half of the 1980s. At Inco the sensor-studded automated smelter could be run by one person per shift, aided by a two-way radio and a microcomputer taking meter readings.[9] By the mid-1980s, machine-vision was also common in Canadian lumber mills and routinely integrated with process-control software and various pieces of computerized machinery for automatically sawing, planing, sorting, and grading lumber. At this stage, too, the focus was entirely on technology, with visions of "lights-out," totally automated factories and mills.

Just-in-time (JIT) manufacturing signalled the shift of focus from capital-intensive technology innovations to organizational innovations. With "just-in-time," computer-integrated manufacturing began to be joined more closely with management information systems. Production was increasingly planned according to intelligence from management's information loop (from market through manufacturers to suppliers), with parts being delivered to factories only when needed on the line. At first the factory was the focus of just-in-time manufacturing, and the object was to decrease inventories. The related purpose was to create a JIT workforce of part-time, temporary, "contingent" workers whose shifts and workweeks, especially in supplier companies, were tailored to the exact production needs of the assembly plant. Subsequently the focus has shifted more towards the market, with just-in-time production geared more closely to QR market-response systems.

Lean production is another organizational technology, this one imported from Japan. It spread through North America in the early 1990s after an industry bestseller called *The Machine That Changed the World* broadcast the results of a study from the Massachusetts Institute of Technology showing that Japanese automakers were able to produce cars with roughly half of all inputs used by their U.S. competitors, and they achieved this through a management strategy aimed not just at reducing inventories, but at reducing production costs in every way, every day. As one observer puts it: "Lean production is lean because it uses less of everything compared with mass production—half the human effort in the factory, half the manufacturing space, half the investment in tools, half the engineering hours to develop a new product in half the time. Also it requires keeping far less than half the needed inventory on site."[10]

Lean production extends the time priorities associated with just-in-time into a governing management principle, to the point that at least one U.S. company, Motorola, has redefined overhead as a function of product-cycle times.[11] As well, new "activity-based" costing systems evaluate (and measure) "period costs" associated with different activities in the production process.[12] Worker-involvement projects such as "quality circles" and "production teams" are all geared to enlisting workers' collaboration in shaving unnecessary "non-productive" time out of daily work so that every move and every second of every minute form "value-added" production time.[13]

As the priorities of time drive production, they are also redefining it in a host of ways. For example, everything—even in the service sector's provision of services—is defined in terms of "flow," with it all oriented to maximizing flow and minimizing bottlenecks. One effect in manufacturing is the increased focus on "manufacturability" (for maximum speed of production, as well as least-cost materials and parts) in engineering design, which is built into related Computer-Assisted Engineering (CASE) systems. This focus biases production towards clip-on parts replacing screws and bolts; and a clip-on contingent workforce, including an increased reliance on contracting out.

As perhaps the final phase of lean production, or the half-way stage to "agility," production systems as a whole are redefined as an integrated time-and-motion process. Production methods are standardized or "commonized" between suppliers and main contractors such as General Motors.[14] Equally, they are rationalized around "integrated scheduling" systems. In launching a conversion to this system in early 1993, General Motors planned to replace all of its eighty-six separate mainframes and software systems with one common system, which would integrate information on everything from engineering specs and parts numbers to sales and marketing data in one "computer and business language."[15]

Agile or "flexible" manufacturing, the final step of integration, reaps the harvest of the others by pushing the scale of production from individual factories and supplier workshops to a network of production sites networked through a customized laneway of the information highway. Here the Agility Forum is promoting a vast interindustry project called the Factory America network, which, according to one of its principals, is intended to function on the lines of distributed data-processing.[16]

Instead of data-processing work being distributed among a number of computer systems, the idea here is to distribute manufacturing workloads to keep the entire system of corporate production facilities operating at full capacity and to maximize economies of scope, scale, and speed. The work will go to a number of corporate and supplier factories and other worksites, where multifunction, crosstrained workers using commonized procedures and standardized parts will take on production orders contracted out by virtual corporations and enterprises cruising the global information network for advantages in time and space and labour. With agile manufacturing, what had been fixed and unique production facilities take on the look of an infinitely reconfigurable Lego set, with "plug-compatible" interchangeable parts, "plug-compatible" manufacturing subsystems, and even "plug-compatible" production units associated with remote and local suppliers, assembly plants, and other facilities.

This Lego-like approach features not only common and interchangeable parts, but also common interchangeable people—or what one skilled tradesman refers to as "cookie-cutter" people and "cookie-cutter" production.[17] Furthermore, with the networking phase associated with the information highway, this standardizing approach is being extended from production to distribution, warehousing, and sales. Furthermore, this "plug-compatible" labour force can readily be de-institutionalized from traditional employment bases to become freelance teams or even individualized labour units credentialled through standardized crosstraining. These labour modules also become readily interchangeable through standardized labour regulations, payroll taxes, and wage/compensation rates.

Agility comes into its own in the context of the new global economy. It represents the operations arms of the new virtual corporations and multicorporate "federated" enterprises working the global information highway to make business happen. As such, the steps leading to agility have had the effect of lifting the whole production process, including what it means to be employed, clear of fixed co-ordinates in time and space. Production, and employment, are being lifted into the orbit of twenty-four-hour global information networks, the site of the new economy. At first it was capital flows plying these networks, with financial interests investing here or there for a day, or for an hour's worth of gain. Now the increasingly high-capacity information infrastructure is filling up with production data: long and complex programs for programming numerical-control machines in this workshop, or robots in that assembly plant, or production planning and scheduling instructions for contingent workers at a supplier workshop or distributor.

Given "Moore's Law," the capacity to bring more and more activity into the orbit of networked information systems takes on the aura of a technological imperative, and it seems unstoppable—even unchallengeable. In manufacturing, this means computerization extending further into the craft trades such as tool and die-making. Stereolithography (SLA) represents the latest extension, perhaps as large a step as Numerical Control (NC) represented in the 1950s and 1960s. It involves taking design data generated by computer-aided design and engineering systems and translating the data into a complex series of computer-code instructions that an ultraviolet laser can use to "build" a part out of a bath of liquid plastic. The laser acts as a machine tool, entirely guided by software, not by a skilled machinist. Furthermore, whatever plastic is not "polymerized" into the hard shape of the required part remains liquid—a perfect "liquid asset," so to speak, for the next part to be made by the infinitely reconfigurable moulding machine.

The literature I've seen doesn't specify the amount of computer memory that SLA requires, let alone the carrying capacity of the communications infrastructure required to transmit the instructions from various sites that might generate various parts-design orders to remote or local SLA moulding workshops—which presumably might operate out of basements and people's garages. Without this information, it's impossible to know whether SLA actually increases productivity, or to assess the trade-offs involved. Instead, the literature boasts of time savings derived from the technology by itself: an 80 per cent reduction in the move from concept to production part, and from sixteen work-hours under conventional tooling to one work-hour with SLA.[18]

There's no mention that this technology shuts down one of the last of the old-time crafts. As one toolmaker told me, "The machines take away our labour. They take away the toolmakers."[19] The machines also transfer control over the work into computers, shifting it from craft workshops into engineering departments under the direct gaze and hand of management information systems. Technology historian David Noble has traced the control theme as perhaps the central reason for the "choice" of NC equipment thirty years ago. As his research into the ascendency of computer controls over the simpler, cheaper, and more developed and popular technology of "record-playback" made clear, about the only advantage of NC was that it involved "a shift of control to management."[20] A survey of machine shops using NC technology in the late 1960s confirmed a "dramatic transfer of planning and control from the shop floor to the office."[21] Once this happens, once the digital divide has been created, the possibility of people programming the new technologies to expand the scope for involvement, creativity, and diversity in their work is shut down. Instead they are confined to being operatives of the new technological infrastructures, and even programmed by their monoculture of technical productivity as well.

Meanwhile, Moore's law drives the dynamo of restructuring more and more in this direction, generating more sales for companies in the capital-equipment industry in the core of the economy and substituting more machines for people and more machine intelligence for human intelligence and skill. It all keeps the economy growing, however joblessly. It's all good, double-good, and double-plus good.

There are mitigating effects. For instance, case studies show some trickling down of professional work, from design engineers to technical support staff who work with design-support software and computer graphics in enriched job descriptions. As well, machines seldom fulfil the promise of either the advertisers or their engineer-creators. The disappeared media stories of the 1980s include the reports of robots smashing cars and destroying each other and a baggage-handling system that shredded people's luggage. In other words, for every application of Moore's law, it seems there is another instance of Murphy's law: the continuing reliance on workers and their often extensive tacit knowledge in the still-open oral culture of everyday technological practice in the workplace. Still, the scope of that culture is being steadily diminished as dependency on ever costlier technological systems increases and as highly skilled tradespeople are replaced by less-skilled technical staff, sometimes on a part-time basis.

As the case studies indicate, there has been, along with the polarization of work, a consistent hollowing-out of local control and human involvement as computerization has transformed everything from primary processing to secondary manufacturing. The Fordist work model associated with routine mass production and bureaucratic work might be giving way to a minority model of upscale flexible specialization, which some commentators call "Toyotism,"[22] but only for a privileged few. Meanwhile, a more computer-controlled, computer-simplified model seems to be emerging for what could be a marginalized majority.

Perhaps the most important theme of the case studies in processing and manufacturing is the transfer of control from the hands-on workers and their working environment into computer systems, or from the shop and factory floor into professional offices. Often there is a struggle for control, which management usually wins, at times with the aid of outside arbitrators.

There is also a struggle over access to training, not to basic training, but to in-depth training that would go beyond merely machine operations—for workers to be able to program the new technologies and not simply operate them. The struggles bear on the polarization of work into a core of full-time professional and technical workers, often routinely working overtime, plus a periphery of often permanent part-timers with no job security or benefits. Finally, there has been unemployment, though this sometimes occurs indirectly through the closure of plants with obsolete, often second-hand technology associated with the "miniature-replica model" of twentieth-century industrial production, and sometimes through jobless growth, privatizing work to prosumers and a reversal of contracting-out so that new networked equipment can operate at full capacity.

Case-Study Evidence

Resources: In general, the more fluid a production process, the more it has been automated, and in a way that decimates the employment. So today, pulp mills, refineries, and smelters are run by a small fraction of the staff levels of the 1960s and 1970s.

At Inco in Sudbury, employment shrank from 18,000 in 1971 to 6,500 in 1987. Between 1988 and 1994, Inco's employment dropped by a further 14 per cent. In the aluminum industry, a Quebec study found a massive decline in production-worker employment accompanying a complete reorganization of work at the integration stage of automation. Loss of control was also a major issue, as control over the smelting operations was drawn into computer systems. Furthermore, "The operators do not have access to the computers because they are located in the supervisors' office."[23]

In forestry, a study of four B.C. sawmills documented a substantial drop in employment, from 1,435 to 891 after a modernization program, with declines in every category up to the level of skilled mechanics; although most of the loss was among the production workers.[24] In pulp and paper, a 1994 study predicted that employment would drop from 72,000 people in 1991 to 52,000 by the end of the decade.[25]

The construction industry has been affected too, especially with the shift towards modular prefab construction, plus computer-aided design. A Quebec case study found that "three per cent of jobs in the industry disappear permanently every year," and noted "a downgrading of carpenter-joiners' skills as skilled tradesmen."[26]

In fish plants, automation, which began with packing and packaging, moved into the main part of the plant, with automatic gutting and washing machines. The West Coast fishery saw a 40 per cent increase in the tonnage of salmon processed and canned and a 23 per cent decrease in seasonal employment after B.C. Packers built a highly automated fish-processing plant in Prince Rupert, sucking in over 50 per cent of the coastal salmon business and throwing the smaller, older local plants out of business.[27]

In a sudden and sweeping move to cut costs in an era of depressed commodity prices, grain companies have transformed an almost craft-like activity run by teams of grain-handlers operating switches and levers throughout the grain elevators themselves into a cybernetic system completely managed from a computer control room. In Prince Rupert the machinery in a highly automated grain terminal built by a consortium of major grain companies cleans, dries, elevates, blends, stores, and moves grain out for loading into ships' holds. Boasting the highest capacity of any grain terminal in the world, the site has increased its throughput to the point of handling 50 per cent of the country's grain exports. Its staff of sixty-five over three shifts is less than what was required to run the smaller old facility for one shift only.[28] Here too there was a struggle for control in the computer room, which back-to-work legislation resolved in management's favour.

Management jobs are lost. But, more importantly, especially in the "country" elevators dotting the prairie landscape, the work of management is being changed from ad hoc responses to local communities and general "rule of thumb" management into something more technical, defined and driven by the central computer system. As one manager explained: "The computerized system has standardized operations in all company elevators, and head office can now monitor them. This speeds up communication with head office, but reduces my individuality in performing procedures."[29] In keeping with the general move towards credentials, the U.S.-based multinational Cargill has begun hiring its elevator managers from among university graduates with systems and commerce backgrounds as well as agricultural economics.

The garment industry: Canada's textile and garment industry is the third-largest source of manufacturing employment. While it's a ghetto, it is also an employment niche, especially for immigrant women. But the niche is highly vulnerable to the technologies transforming this industry.

Computerization has become a prevalent corporate response to free-trade-related clothing-import competition as well as deregulation and the drying up of subsidies to support a Canadian clothing industry. By the mid-1980s, the E-ton sewing assembly line from Sweden had become fairly common in the Canadian garment industry. It has more than fifty individually engineered reprogrammable work stations with a monorail-type line for moving materials from one station to the next, each one equipped with sensor-controlled gates that cause the material to stop or to pass by, depending on the instructions keyed into the system.

At each work station the individual operator receives the garment off the monorail conveyor with one hand, feeds it through the sewing operation, and loads it back onto the conveyor unit with the other hand. At each work station the most sophisticated E-ton 2001 system incorporates a computer-monitoring device that can relay information on the operator's work both for materials management and for calculating wages. After the introduction of an E-ton automated sewing system in a Toronto blouse factory in the mid-1980s, the workforce declined from ninety-eight to fifty-seven.[30]

By the early 1990s the focus had shifted from production to distribution and marketing. In the United States, TC^2 is a U.S. government-supported consortium of 250 clothing and equipment manufacturers, 16 "off-shore organizations," and clothing industry associations, plus academic institutions. The group has combined reconfigurable computer-integrated clothing manufacturing technology with quick-response merchandising and distribution to create a new industry standard for agile clothing production that amounts to virtual clothing or "apparel on demand."[31] The customized Levi Strauss jeans available at the company store in Toronto's Yorkdale Shopping Centre are an example of this in action. Through a multiple choice of over four thousand possible combinations of hip, waist, inseam, and rise, customers can order their own made-to-measure jeans via the store's order-processing computer. The store guarantees delivery in three weeks—though it hopes to reduce this to five days.[32]

While the media gave no details about how the work was done, a 1992 women's conference on homework referred to Canadian garment manufacturers that buy cloth manufactured in China, have it cut (with laser cutters) in New York and sewn in El Salvador, and then sell it in Toronto.[33] Increasingly, the garment workers are "homeworkers" in the trend towards self-employed contract workers networked electronically, pioneered by Benetton.

Agile manufacturing enables a radical de-institutionalization of work once the work itself has been computerized to McJob-like status—that is, when its operations are determined and controlled by the computer system. In the garment industry at least, this is resulting in more women working out of their homes, both nationally and internationally, while garment factories are being closed in countries such as Mexico and Nicaragua.[34]

A survey of working conditions among garment trade homeworkers in Canada found that twenty-one out of thirty were being paid less than the minimum wage, and twenty-seven out of thirty were experiencing health problems: physical allergies from the dust plus mental stress from the piecework production demands. They were also working in poor ergonomic conditions as they carved a work area out of a space designed for sleeping, eating, and living. They worked long hours—averaging forty-six hours a week, with this rising to seventy hours during busy periods. Labour law excludes them from overtime pay provisions, and many relied on other family members—child labour—to get the work done in time. Most had resorted to homework because they couldn't find or afford child care. As well, only one woman in the survey could converse in English, which the researchers interpreted as racism reinforcing sexism to marginalize immigrant women.[35]

Railways: Canadian National Railways implemented a system called TRACS for keeping watch on all railway cars and their freight loads. In the mid-1980s this system was integrated with the Yard Inventory System in the various repair yards in which engines and cars were maintained and repaired across the country. Since then the system has been integrated with the management of hump yards, where cars of similar freight and destination are assembled into trains and others are disassembled as they go separately to final destinations. Ostensibly this computerization has been driven by the need to cut costs, especially in the face of competition from trucking. But in the way that it is being done, in taking the initiative and control out of repair and yard crews and concentrating it in office computers, the system reduces many workers to "pin pullers, that's all."[36]

The shift of work from the yard into the yard office has led to a combination of formerly fragmented separate jobs into multiskilled, multifunction jobs such as a "servo-centre control clerk." Still, a report on CN continues, "We are extremely doubtful if any of the present unionized rail workers will reap the benefits of this skill enlargement in more than just the short term, since we expect the majority of present positions to be subject to absorption by the on-line computer systems themselves, with a minority remaining in 'monitoring' jobs."

Since 1990 CN has dropped twelve thousand jobs from its operations—a "brutal downsizing" that some analysts have called a prelude to privatization, but which technological restructuring can only have facilitated.[37] Similarly, in a post-free-trade restructuring, Canadian Pacific reduced its labour force by 54 per cent, from 85,400 in 1988 to 39,300 people by 1994.[38] In fall 1995 the company announced plans to cut a further 1,450 jobs in freight-management and support staff in a major corporate reorganization coinciding with a head-office move from Montreal to Calgary.

Telecommunications: Northern Telecom is the wholly Canadian-owned former subsidiary of the U.S. telecommunications giant AT&T. The manufacturing adjunct of Bell Canada Enterprises, it represents Canada's key heartland industry in the digital new economy. Yet in another post-free-trade continental restructuring, Northern closed plants in Quebec and Ontario as well as some facilities in the United States. The Canadian Auto Workers represented 5,000 Northern workers in 1985; ten years later the workforce was down by half, to 2,500 workers.[39] By 1993 its U.S. assets represented 40 per cent, compared to Canadian assets at 24 per cent—down from 45 per cent ten years earlier.[40]

Overall employment in Canada's communications and other electronics equipment industries dropped by 20 per cent between 1986 and 1991—that is, during the postrecession recovery. Women's employment dropped by 24 per cent, compared to 18.5 per cent for men. Within that, it seems that women are being reconcentrated in ghetto-like positions at the lowest end of wages and involvement.

Most of the manufacturing work in telecommunications is done by machines. On the circuit-board floor in Brampton, Ontario, a robot scans and identifies the board coming into its automatic assembly substation. It then pulls the assembly sequence required for such a board from its machine memory and communicates instructions to a second robot, telling it which parts to pick out of the pigeonholes arrayed beside it. As the second robot hands each part in turn to the first, machine vision helps guide the placing of the part into its appointed spot on the board before another piece of equipment automatically pushes and clenches the part into place. In another work station, automated machinery bonds the parts together and coats them. Another substation then tests them for defects. And on it goes.

Workers who used to operate machinery in the production process now tend machines, monitoring the work done by the various robots and feeding them parts out of supply bins associated with "just-in-time manufacturing." Specially coloured cards indicate low stocks and trigger an automatic reorder.

One of the few exceptions involves putting custom-ordered parts onto circuit boards. But where this job used to be the work of one person at a workbench, it is now fragmented into eight standardized tasks organized along a variation of the assembly line called a slide line. Workers—the majority of them at the Brampton plant are now women—pull a plastic carton containing the work along a countertop into their work station, where they perform one module of operations in the circuit board's assembly. Because the workers rotate jobs through all the work stations, management regards the slide line as an example of enriched work and reskilling, as "doing more on the job" and "increasing the variety of tasks."

Workers who can remember when the job was a holistic craft are less impressed. "Do you think I'd rather be putting in eighty parts and building a board, or putting in six transistors eight hours a day? It's degrading!" They can easily differentiate between job enlargement and job enrichment, multitasking and multiskilling. As one said, "So we can rotate on the slide line. Instead of putting six capacitors in, we can now put six resistors in. Big deal!"[41]

In production maintenance, Northern collapsed the jobs of electrician and "machine repairman" to create a new semiprofessional position called electronic technician. But the computer does most of the work, first test-scanning its way through a defective piece of equipment to identify the problem part, then displaying its identity code and location on the computer screen. All the technician does is fetch the part and put it in.

"They don't want tradesmen; they want part-changers," one technician said. As if to drive home the point, the electronic technicians here and in drafting are part of a trend back to night shifts: their work is that divorced from the company's full-time day community—and they are that much more "costlessly replaceable." It is sometimes difficult hiring a skilled tradesperson, particularly one who will work until midnight; it's easy finding someone who can change a lightbulb—especially if there's lots to choose from, and if jobless economic growth keeps people grateful for any work at all.

Skill requirements are reduced in a range of indirect ways: as a host of specialized parts are replaced by a handful of standardized parts; as screws in assembly are replaced by snaps; as several individual tests are collapsed into a computerized multitesting unit; as production cycles are pared and waiting times eliminated; and as computer diagnostics reduce defects to the brink of extinction. (This doesn't mean that the product is necessarily any more durable or otherwise better. It just means that it's been produced exactly according to the computer-designed specifications. There's practically no room for human intervention anyway.)

The Northern case study noted, "The staff associated with transporting workpieces, preparing schedules, recording data, accounting for parts, tracking work orders and similar activities are no longer required in an automated system." The system does it all.

There is still a need for skilled workers. For instance, in one year the company hired nearly a thousand engineers, computer analysts, scientists, and technologists while dropping sixteen hundred production workers. However, many of the new skilled workers are doing little more than "babysitting" robotized equipment, once the start-up bugs have been eliminated. Their computer skills are needed only to recalibrate the work station when a new design order comes down the line. As well, the need for new skilled workers is being outstripped by the skilled machines' ability to simplify and automate existing work. In the way the technology and work are being organized, more and more people are being trapped behind the new silicon curtain, in computer-simplified, computer-controlled jobs, working for the technology, not with it. They're also being further marginalized in shift work and part-time jobs.

Northern Telecom was one of the first manufacturing companies to move towards a two-tiered labour force, by hiring "temps" for some of its smaller-batch assembly jobs. Significantly, most of these "permanent" part-time workers are women.

For its office work Northern Telecom has bought expert-system software to transform strategic planning from a paper-clogged human-communications process taking six months to a standardized technical operation requiring less than six weeks. Similarly, integrated information systems allow the person designing a part to automatically generate an electronic order for it, bypassing the clerical worker who would otherwise generate a purchase order and circulate it through the mail. In the London telephone plant, clerical ranks were cut by 60 per cent between 1981 and 1986, while the electronic technicians increased their share of employment to 32 per cent and professionals boosted their share to 28 per cent.

At Brampton clerical workers are increasingly finding themselves divorced from the corporate community and narrowly confined to a ghetto of computer subroutines. They are taught to press special-purpose keys, but are not informed about the programmed meaning of these keys. They are told, "It's none of your business. You don't need to know."[42]

Similarly, restrictive codes and passwords through the information system serve to lock employees into narrow job functions more completely than when old-fashioned line supervisors did so with overt authoritarianism. One of the computer clerks said, "They isolate information or segment it," adding, "No one gets the whole pie except management and the confidentials [the company's name for professionals and other designated, privileged people]. They have all the commands."

Another clerk said: "The password system restricts you. Like, you are given a key to the first drawer, and someone else is given the key to the second drawer. That is how they isolate you. . . . I had more decision-making before. Now they tell you what to do and you follow."

Ursula Franklin calls it the new "headless tyranny." Yet because people's exclusion is built in, the system seems neutral if not natural. The system proves impossible to challenge, because there's no one there when the door won't open in response to the swipe of your magnetic I.D. card, when a computer file stays closed, or network access is denied. No overt authority figure stands guard, ready to say no. There is only the blank-eyed computer screen, and human rage, impotence, and alienation conveniently do not compute.

Other restructuring choices deepen this computer-enforced schism: the streamlining of work around machine systems to facilitate reliance on both temporary and shift workers; the increased emphasis on formal credentials over experience and tacit knowledge; and the unequal distribution of training. Again and again at Brampton, production workers found themselves on one-day or two-day training courses while professional staff went off for two weeks of theory and hands-on practice. One worker said bluntly that the engineers "make sure they get all the good training, and we're stuck sucking the hind tit."

A study at a plant codenamed Quality Electronics confirmed the drastic drop in women's share of employment: women formed 70 per cent of the factory workforce in the 1970s, but by 1990 their representation had dropped to 52 per cent.

Tellingly, none of the women had less than seven years' seniority. As the study's researcher, Karen Hadley, discovered, "While there had been several middle-graded technicians in the plant just prior to my research (in 1990)—young women who had graduated from college technology programs—they had been laid off because of their low seniority. This is a phenomenon found in many manufacturing workplaces (e.g. steel, aircraft) where women were finally breaking into 'non-traditional' jobs. They are the first to be laid off as plants downsize."[43]

Women are losing in the struggle for diminishing training opportunities, and they are losing in the shift from experience to credentials for promotion. With the women carrying 90 per cent of the load for child-care and housework, Hadley reported, "Women's double day prevented virtually all of them from qualifying for jobs [as technicians]." The bias went beyond gender to include race and ethnicity. The many South Asian women at the plant were "disadvantaged" by a training model that was heavily text-based, did not "reflect what they are capable of," and penalized them for their ethnicity.[44]

The study documented more hollowing-out of middle-range jobs. Production jobs were being reclassified from grades five and six down to grade four, and some of the more highly skilled work was being relocated to a sister plant in the United States. While before there had been 281 separate classifications in production, now there were 80; where there had been 32 classifications for technicians, now there were 12. The result was a steady intensification of people's jobs. "By expanding the tasks workers can be called upon to do, and controlling where they can be placed, management gains a means of eliminating any idle time in the production process."[45]

The pace was pushing many workers into early retirement at fifty-five. And once the dust settled, much of the highly touted "multi-skilled" new work emerged as more "standardized jobs" that were "unbearably monotonous." As Hadley put it, "Most assembly workers at QE felt that the process of standardization and cross-training/multi-functionality had both diluted the skill content of their jobs and intensified the work pace."[46]

Automobiles: A Canadian Auto Workers' survey done just prior to the Free Trade Agreement in the late 1980s found that companies were buying automated equipment instead of recalling workers laid off during the recession of the early 1980s. Some 60 per cent of the plants in the CAW study reported new computer equipment: robots, computer-controlled machine tools, and automated material-handling machinery. Half of the plants acquired just-in-time process-control technology, four-fifths reported statistical process-control technology, and one-fifth got new tool-and-die change techniques.[47]

The survey found a high coincidence between the introduction of new technology and job loss or displacement. For instance, Excello Colonial permanently laid off six lathe operators after installing numerically controlled lathes. When Kenworth Truck introduced automated welding and painting equipment, the number of welders dropped from fifty to sixty down to fifteen to twenty. The number of painters dropped from sixty to thirty.

While automation eliminated a lot of routine, repetitive jobs, it also downgraded many formerly challenging jobs into something close to drudgery. While a third of the respondents said technology had eliminated or improved "unpopular" jobs, a quarter reported that it had cut out "popular" jobs as well. Interpreting the shifts is complicated by the fact that many automobile companies are contracting out the maintenance and repair of the new equipment, and "out-sourcing" subassembly work—for example, engine parts and car seats—to Third World countries or depressed low-wage regions of the Americas.

There are many examples of deskilling. Assembly has been simplified into plugging together a diminished number of more sophisticated electronic parts or modules; doors and other parts that used to be welded are being clipped together or replaced by plastic-metal hybrids made by automated injection or stereolithographic moulding. In the shop, automated lathes and other computer numerically controlled (CNC) tools systematically degrade machining from a craft to little more than babysitting the machine.

One worker said, "The guys they brought in for the CNC machine had no tool experience. . . . They got hired in at $8 an hour. . . . They just put the tape in and sit and watch the machine."

Most of the reskilling is in the form of multitasking, although even this is seen as an improvement. Compared to operating the same piece of equipment on a line all day, monitoring and making small adjustments on several pieces of automated equipment in an integrated work station represent an escalation of skill and job satisfaction. Similarly, at the time of the study the task of operating a computer and high-tech equipment still carried a certain mystique, along with the challenge of dealing with a system that still had bugs to be worked out of it. But as time passed, and the system was running smoothly, perspectives could change. As one worker in the auto study observed, "There's nothing worse than a line that's running perfect. It's a boring job."

Control emerged as a major issue. Partly, as management's engineering, drafting, and numerical-control software took over, this involved the loss of leverage previously gained through machinists' craft skills and related working knowledge. Partly too, this transferred knowledge gave new leverage to management to both boost production standards and enforce them through sensor-feedback and other forms of performance monitoring.

In a Ford glass factory, a worker reported: "If your count is 1,000, they say they'd like 1,100. We say we can only produce 1,000. They ask for more and we say we can't produce more, and they say, 'Fine,' and they end up bringing in the automation."

The workers call it "management by stress."

The survey showed that workers' attitudes were split on the issue of control: 81 per cent felt they had been given more responsibility; and 82 per cent felt that management now had more control. But 84 per cent of the workers felt that they were under more stress.

Aerospace: Like telecommunications, aerospace is part of the high-tech heartland of the new economy. Here too, much of the creativity behind computerization is being channelled into simply automating work that was already being done and bringing it under tighter computer-management control.

At the time of a Canadian Auto Workers' study at de Havilland Aircraft in 1988, computer-aided design systems in the office were being linked with data bases in the factory for increasingly computer-integrated manufacturing. The company was buying more numerical-control machines and hiring more programmers to transform more work into computer-controllable tasks. Metal parts moulded by skilled tradespeople were being replaced by plastic parts, which could be made automatically through computer-controlled, laser-operated plastic-moulding equipment. Fabricated parts, which had previously been stamped and welded by workers, were now being replaced by computer-machined parts. The company boasted a threefold productivity gain, though without taking the cost of buying and running the computerized equipment fully into account.[48]

Most of the job effects were indirect. For instance, whereas the company used to contract out 90 per cent of its machining work, that job was now being brought back inside. Automated riveters that drill a hole, countersink, put in the rivet, and pack it can do twenty rivets a minute, while manual rivets had taken three minutes each.[49] The change had not resulted in layoffs, partly because the work tended to be short-run batch operations. But formerly separate machining operations were being integrated and downgraded as more and more machining work was computerized and more of the programming work removed from the factory floor.

The de Havilland machinists have a collective agreement that commits the company to keep programming on the floor, and they were actively resisting this change. (By contrast, the Canadian subsidiary of McDonnell Douglas expressly proscribes programming at the operational carousel.) The autoworkers' report describes the operative clause at de Havilland as resulting from the union's "considerable foresight." But it could perhaps also be understood as partly the legacy of de Havilland's workplace culture as a medium-scale public (Crown) corporation, less colonized by the controlling, cost-cutting imperatives associated with U.S. big business—at least, until it was taken over by Boeing in the 1980s.

The CAW report is peppered with contrasts: machinists transformed into "push-button" operators, while some new office technologies upgraded jobs. One worker said, "I have more responsibility now." Another stated, "Now, all we do is add information to the computer, and look up information on it."[50]

Clearly, computerization was once again centralizing control, according to the report. "CNC gives management more control over the planning aspects of machining work and makes the operations of production more visible to management. In that sense, production is more controlled and decisions are moved further upstream."[51] And workers were again losing what control they had previously enjoyed under the old craft-centred, more holistic approach to work. In the switch to the new computerized equipment, the skilled workers weren't being taught to program the new machines or to work with them as creatively and autonomously as they had with the old machine tools. A two-day course on CNC that a limited number of them were able to take only gave instruction on how to operate the new equipment, stopping far short of letting them exercise their contract rights and really program the machines. "It's not a case of 'do this and then you do that,'" one of them said. "Repeatedly, operators refer to being trained by other operators.... 'One guy had taken a programming course, so we went to him.' . . . 'Fred had sat with the vendor while the machine was being installed, so he trained me.'"[52]

At McDonnell Douglas Canada (MDCan), Mazak CNC machining centres featured a tool "carousel" of up to eighty interchangeable computer-controlled machine tools. As one manager put it: "The Mazaks are running three shifts, the men are on one shift. They just push the button and go home." The machines were running twenty to thirty hours non-stop, machining a complete "nest" of parts. Dovetailing with this, MDCan had also become a market for the parent company's Unigraphics semi-automated design-for-manufacturing programming system. With it, "A programmer can draw parts, nest parts, create fixtures, do lay-outs, plan the job, determine the tooling, try-out and verify the job—all on screen."[53] As a result the number of machinists dropped to roughly a third of those used under conventional machining. Programming had clearly become the responsibility of office-based programmers. And machinists were being replaced by new multitask, multiskilled "work-unit technicians," at the second-highest pay level. But in a pattern that seems typical for production workers fitting a "Toyota" variation on the post-Fordist work model, these relatively well-paid, full-time workers were regularly expected to work overtime. The company's training manual made clear that shifts would be staggered with a four-hour buffer "to allow overtime to be worked when necessary to make the required parts or to complete the required assembly."[54]

Again there were complaints about not enough training, and not enough depth. One worker sounded a familiar note: "They just showed me how to load the part and push the button." Complaints about stress, the loss of down-time, discretionary, or "buffer" time, and "increased production with fewer workers" were common. "With this JIT system," one worker said, "you're always under pressure to produce." Another said, "If there's slack, they'll fill it. If you get ahead, they'll give you more work."[55]

Job losses were indirect, including work in the office. Job vacancies were regularly going unfilled. Work was also being transferred out of the bargaining unit, often into the hands of non-unionized temps. In 1981 MDCan had between 700 and 750 unionized office workers and 800 "excluded" staff. By 1988 the union's ranks had dropped to 430, while the excluded staff levels remained unchanged. Sophisticated information systems, such as material-resource planning systems, automated stock-locating systems, and production information-control systems were digitizing office work everywhere. For example, the production information system collects attendance information, labour distribution, and hours paid for all departments, automatically balances labour distribution with hours paid, and feeds this to payroll. As well, jobs are sidestepped through electronic integration between the company and its clients and suppliers. With information highway connections, these links could become more and more common, shrinking employment further and centralizing more creativity and control in the company's head office.

Office workers complained about inadequate "introductory only" training and waged a similar struggle for training opportunities, finding discrimination both against unionized staff and, within that group, against women.[56]

TQM and Integrating People into the Corporate Vision

In agile manufacturing the agile interchangeable people required to work within these systems must become agile jugglers. Or, perhaps more accurately, they become the juggling balls in the hands of management on the quick response/flexible manufacturing playing field.

The quality-focused team-management concepts associated with lean production are critical in this regard—especially when quality is defined in Japanese-management terms as "performance to specifications." TQM programs adjust people to become not just extensions of computer-operating systems but intelligent parts of that system, actively assisting in fine-tuning it. Through corporate "cultural training" the system hooks the workers into management's technological imperative of cheaper and faster equals better. And people's often unacknowledged working knowledge—their "tacit knowledge"—is continually appropriated into the computer system.

The corporate-culture aspect is obvious in the PR component of quality circles and total quality management. When I toured Northern Telecom's Brampton factory in 1988, multicoloured banners throughout the factory proclaimed "The Heat Is On." The slogan is the title of a half-hour documentary the company commissioned outlining the dangers of off-shore competition and the merits of emulating the Japanese style of labour-management teamwork in the interests of sharpening Northern's competitive edge. To narrate the film the company hired Walter Cronkite, that fatherly news anchor who articulated public opinion to America for decades. It also sponsored an essay contest on productivity and competitiveness, with winners sent on a tour of Korean auto plants then featured as speakers for productivity testimonials on their return.

From the management point of view this is a post-Taylorist, inclusive industrial culture. As one Northern manager put it, "We're not just hiring people from the neck down." Another said, "You manage now by involving your people, by mutually working things out."

At the same time as workers are being invited in, however, access codes built into work-station computers restrict the scope of their work, and the company's new magnetic-strip identification badges restrict their mobility around the facility to only those areas strictly pertaining to their work.

Workers complain, "They keep everybody in the dark." Management replies, "We tell them as much as we think will be beneficial to them and healthy without raising hopes or fears." Bruce Roberts of the Canadian Auto Workers says that if workers want to check on something that's happening on the line next to them, they can't get access to it. "That's why this Information Highway is such a hoax, for us."

Yet worker-involvement programs have moved into practically every line of work since the late 1980s, and teamwork is the message on its telescreen. Among the range of manufacturing plants represented by the Canadian Auto Workers, Dave Robertson, director of work organization and training, estimates that this cultural training—including seminars on corporate marketing and global competition—accounts for the bulk of "training" that workers are now receiving on the job. A national survey published by Queen's University's Industrial Relations Centre also found that "about one-half of respondents" offered what they called "social" or "cultural" training—for instance, in leadership, communications, group decision-making, and team building.[57] Another 26 per cent reported training programs on health and safety, while another 18 per cent reported "orientation" programs. "Less frequently reported" was "training more closely linked to developing human capital—such as computer training."

At Northern the program is called "Team Excellence." At the CAMI automotive plant at Ingersoll, Ontario, morning exercise sessions, mandatory for non-unionized office staff, typically end with a collective cheer for the plant. Posters throughout the assembly plant proclaim, "CAMI: A total team effort." Banners read, "Safety on the job is a team effort."

One "Corporate Communications" event involved about a thousand employees in a tent outside the Ford Motor Company's plant in Oakville, Ontario. The company offered videos about small-town football teams, and then a management official got up and congratulated the workers on their success at improving the quality and on cutting down the costs of producing the new Tempo and Topax cars. But, he said, there was a "new reality" out there. An industry shakedown was coming, and if each of them wanted to be a winner, "You're going to have to produce a lot more with a lot less." He compared Oakville's productivity stats with figures from a Kansas City plant producing the same products. Kansas City was ahead. "Look, despite this," he said, "I have a lot of confidence that the people in this tent have the ability to be winners, and that we can beat Kansas City." This brought a cheer from the crowd.

The corporate spokesman announced that he had managed to recruit a top manager from Kansas City to come to Oakville. The crowd cheered again. The manager of the Ford plant came up and told the audience that he had a lot of confidence in them too. He said, "Watch out, Kansas City. Here comes Oakville!" This time, "The whole tent erupted and people threw their caps in the air and they started shouting. . . . And then, as if out of nowhere, the tent filled up with the sound of very upbeat rock music—really pumping people up."

According to political scientist Don Wells, who reported on this event, "These kinds of communication days were taking place in every Ford assembly plant in North America."[58]

A religious revival meeting or a sports pep rally? Either way, the effect is the same. As Noam Chomsky observes, team-competitiveness hype is "a way of building up irrational attitudes of submission to authority, and group cohesion."[59]

What's more, this phenomenon substitutes another "enemy"—sometimes one with racial coloration—for management in traditional labour-management affairs. It also offers a welcome escape from a grunt identity associated with being a "mere" manual worker. This is especially appealing for women who have found themselves rigorously confined to manual jobs in factories while the mental work is reserved for men.

As one woman in the Quality Electronics study said, "Before we were just the lower part of the company that nobody really cared about. . . . Now [the women] realize that their job is the most important job because without the people doing the assembling and putting the circuit boards together properly, [QE] wouldn't have customers."[60]

To be called a "co-manager" and treated as a teammate by a boss walking around in blue jeans is almost irresistible. It also speaks to a heretofore unacknowledged truth: workers regularly bring much more than manual "skills" to the job and do much more than execute management's procedures. This truth was simply never formally recognized, before TQM came along.

TQM and Tacit Knowledge

Working knowledge, tacit knowledge: It is the invisible knowledge, the know-how filling the gap between the theory of prescribed work and the reality of getting it done. It occupies the space between the assumption of expert knowledge and of clearly prescribable practice and the truth of local, contextual, and personal knowledge associated with getting things done in the here and now.[61] It is the informal reality behind what Ursula Franklin calls the "myth that the only real knowledge is certified knowledge."[62] It is hardly ever acknowledged (except sentimentally, through National Secretary's Day, for example) and never properly rewarded. But it's essential to making the economy actually work, everywhere from offices to factories to nuclear weapons laboratories.

In the early days of the telephone, mechanics and telephone operators were strategic agents of technological development, helping to articulate what telephony could be in terms of both telecommunications infrastructure and public service. According to one account, "Technological change occurred on an almost daily basis in the Central Office as these mechanics went about improving the equipment. In fact the apparatus of the Toronto Central Office was totally changed five times between 1880 and 1914."[63]

An interview with Grace Witty, a switchboard operator in Northern Ontario, is full of revealing details about the knowledge she brought to the job, and about how much she did with that knowledge. "The linemen came from Sudbury. I was the one who would tell them about the trouble on the lines. . . . I liked the managers I had. I taught one . . . the ropes before she took over. Bell Telephone paid me twenty-five cents an hour. Weren't they generous?"[64] She also extended the meaning of telephone service. Once a local hockey team persuaded her to announce an upcoming hockey game. She plugged everyone in at once, threw all the keys, and made the announcement, turning the telephone into a broadcast as well as a communications medium, a feature that in some places—in Brantford, Ontario, briefly, and also Paris and Budapest—was developed as an integral part of the phone service.

Taking this kind of knowledge seriously, insisting on bringing it into the picture, Ursula Franklin describes telephone operators of the early twentieth century as what "today would be called 'product-development engineers.'"

Franklin states: "During this phase, in which various applications of telephone and telegraph communication were developed and tested, the operators were the central participants in the experiments. . . . The operator's role was that of an operating and trouble-shooting engineer as well as that of a facilitator."[65] In other words, without the tacit knowledge, and reliability, of the operators in their daily contact with customers, the telephone industry would not have succeeded as it did.[66]

A study of tacit knowledge in the development and handling of nuclear weapons also stresses the importance of involved working knowledge, and not just in the early stages of technological development. In fact, the authors argue that this knowledge is still so critical that its lapsing and possible disappearance as an effect of continued nuclear test-ban treaties might even force an end to the age of nuclear bombs.[67]

In the Quality Electronics study, the women knew that they made the place run. As one with twenty-six years' seniority said, "You just do. You just go along. I can't explain it but you do keep up with it. We just have the skills and we don't get acknowledged for it."[68]

Another said: "There have been times when the five women in my area have solved a major problem before it became catastrophic and it's been brushed by the wayside. 'Well, that's part of your job, dear.' Whereas if it had been one of the male engineers that had solved the problem, it's like, 'give this man a raise, give him a corner office with a window!'"[69]

Several women talked of "natural" work teams. One of them said that the women in her area had "always worked as a team" to make the jobs "easier for each other, informally."[70]

Formal "Team Excellence," quality-control programs, and other co-management strategies are simply recognizing the wealth of knowledge that has always existed in the open, oral, living culture of the workplace. But they are doing more than that: they are taking that knowledge over. The programs don't exist to honour the knowledge, except in rhetoric, and let people build on it. They exist mostly to appropriate it, to "black box" the knowledge into computer systems.[71]

Much of the training associated with quality-control and team-excellence programs involves teaching traditional management skills associated with time-motion studies, flow charts, and related planning. In "quality circles," workers take it from there, finding new ways to reduce waste of motion or more ways of paring the production process down to pure "value-added" production.

But, as Karen Hadley points out, "They are all techniques used to generate data on the work process so that operations at the point of production can be simplified, and productivity continuously increased."[72]

In other words, the "training" associated with quality control and team excellence is geared towards a reverse flow of knowledge: from workers into the computer system controlled by management. It robs the workers of a traditional power base.

The training involved is almost a pure case of Orwellian Newspeak: the language used has been so re-engineered out of its original cultural context that words can be accepted as meaning their opposites. In *1984* Orwell showed how "freedom" and "slavery" came to be interchangeable terms. In the 1990s the situation of people becoming more tied into computer systems and their operating priorities equals "empowerment."

This is the con, the cheat of total-quality management and team-excellence programs: because management interprets quality narrowly, to mean optimizing the performance of the production system *as system*. Excellence doesn't mean producing a more durable or environmentally responsible product, and it doesn't mean providing better service. The companies define excellence largely in systems terms: faster, cheaper equals better.

At the CAMI automotive plant, the "taien" suggestion program rewards work teams (with names like "Kaisen Knights" and "Rotation Wrestlers") by giving them microwave ovens for their lunch areas in return for ideas about cutting costs and doing more with less.[73] The program also offers individual workers money and a chance to win cruises and other prizes for their suggestions. One woman was an enthusiastic supporter of the program and thought she was doing everyone a favour when her work team found a way to "kaisen" out the equivalent of a whole job and turn this into a floater position to relieve pressure up and down the line. Then the company got rid of floaters. "And we had just been had big time. Our whole team sort of opened our eyes and thought, 'Oh my God! We've just kaisened a whole person out of our team [to create a floater] . . . and we've essentially screwed ourselves out of that, and thought we were quite clever doing it.'"[74] (Incensed, the workers turned the move into a strike issue, and won a partial reversal.)

Another worker who had contributed a lot of kaisen suggestions in the past complained that he couldn't make money any more: "Everything's been pretty well taiened out."[75] The production system had been about as fine-tuned as it could get, with no slack, no buffer—and tensor bandages all over the place as repetitive strain injury rates had risen, regular long overtime had accumulated, and workers were pressured by co-workers to not take days off.

In the Quality Electronic study, the workers talked about the contradictions. One said, "Management thinks they want you to be a team and take away some of the responsibilities they have until you actually make a decision that will affect whether the line shuts down, or whether you refuse to ship something, and then they don't want you to make a decision anymore."[76]

Another talked about computer-monitoring. "You'll never achieve a team atmosphere as long as we know you've got a computer in your office and at any time you can punch a key, see where we are, how many jobs we've done today, etc. You can only have one or the other."

Not only do the workers participate in making their jobs more standardized and boring, plus more pressured and stressful. But as their tacit knowledge helps management debug the system, many of them also worry that they're setting themselves up for redundancy. As a woman in the Quality Electronic study predicted, "We're involved in changes (improvements in the work process) that will keep Mexicans in their jobs."[77]

But the struggle for control is far from over; the living work environment is much too complex for straightforward predictions. Workers are doing some appropriating of their own, using the team programs as leverage and a site for resistance. Meanwhile, the programs are having a transformative effect, adjusting workers to being operative parts of the global production dynamo.

5

Panopticons and Telework: The New Cybernetics of Labour

It is a technology that is supernatural, acultural, alingual, a technology . . . of binary digits that can saturate the world.
– Marcel Masse

IN 1994 the World Bank floated the idea of security cameras in North American shopping malls being monitored by screen watchers in Africa.[1] From the World Bank's perspective it was a sound proposal that would harness local job creation to debt restructuring, a proposal made technically possible by the high-capacity bandwidth built into the information highway.

If it happens, and it certainly could, it would be a fittingly banal betrayal of the promised Knowledge Society and 500-channel universe. There's no reason why home-based work and other forms of de-institutionalized, demassified employment and self-employment can't flourish on a participatory, inclusive information highway. People should be able to surf the Internet, to cruise the information highway for contacts and contracts, to offer their knowledge, talents, and skills and have them taken up to feed, inform, enrich, and renew our economy and society.

But the results depend on who is designing the structures and who is programming the operating systems. Given how the highway infrastructure is currently being built, and how technological restructuring has been designed and programmed so far, the promise won't be fulfilled. Most people aren't being given real choice or control. More and more people are losing what little control they previously had to computers, and they are coming to be controlled by them. The growing numbers of second-tier contingent workers and of jobs serving as the human extensions or operational parts of computer systems are setting the scene for a rigidly polarized new economy in which only some people will participate freely, fully, and on their own terms. Only some people will be able to use the new global computer networks as an extension of their talents and intelligence, agilely telecommuting to new opportunities for employment and involvement. Many, many others could well be stuck simply at the receiving end of those networks, working as extensions of their logic and operating terms. Many won't even go anywhere to work; they'll wait for the work to be sent to them.

It's too early to predict whether the telework model of computer-defined and computer-controlled tasks dispatched to remote worksites will emerge as the new post-Fordist work model, or at least as a model of work associated with the information highway. But the basic elements of such a model are already in evidence, and they deserve our critical attention. They represent a shift in the distribution of power in our society towards computer systems and those who control them, and a new version of class polarization—here across the digital divide of technological enfranchisement or disenfranchisement, of working with computers or working for them. They also represent a new form of social control: from a human context of industrial relations to an almost entirely cybernetic context.

With the model of work emerging around telework, computer software prescribes every aspect of the work to be done. As people become enclosed in a fully programmed and remote-controlled work environment, they not only work as an extension of the machine; they also begin to think and react solely in its terms as well. To a large extent, these trapped workers can no longer think for themselves, or even weigh matters through consensus. They "turn turtle," as Marshall McLuhan suggested, evoking the image of a turtle turned inside out with its shell on the inside and its organs and flesh exposed on the outside. These people can be programmed or acted upon directly without the protection of private thought and consciousness. "Turtles with soft shells become vicious," McLuhan wrote. "That's our present state."[2]

Digitizing Memory

If you look at re-engineered institutions and workplaces from the perspective of the smoothly networked information and production systems being developed to run them, it's hard not to be impressed—even seduced—by their rationality and technical efficiency. Once you get inside the orbit of the information system, you have almost no choice but to judge it by its operating values: productivity measured in digital terms, according to computer metrics and related accountability. It's also difficult to then judge it in terms of the values and priorities of society—universality, equity, and meeting people's actual needs—except by translating these into equally digital terms: measurable outcomes and flow charts.

To put these massive transformations into context, you have to fill in the blanks in the literature addressing the subject. You must begin with the living context of how things were done *before.*

At Queen's University, computerization followed the normal sequence: from the back office for major accounting and administrative functions out to the front lines of the university's relations with students. Partly this helped justify the growing costs of computer operations and maintenance. Partly too, the technical goal of optimizing the performance of a computerized system inexorably insists that all information needs be brought within the system's orbit. Every step in computerization builds a momentum to computerize more, with multiplying benefits from smaller and smaller increments of investment. Eventually at Queen's this momentum visited itself upon the dean of women, who ran the women's residences. An older woman with many years' experience, she had organized the running of the residences in ways that stressed the personal needs of the women inside.

Then the systems experts came by. They wanted to abolish the night clerk and automate both the turning off of lights and the locking of the front doors. They also wanted to automate the payment of students' residence fees. The suggested changes made eminent sense from a systems-efficiency perspective. They were workable, and they would most likely save money as well. But the changes made no sense to the dean of women.

She explained the situation from the perspective of looking after a community of young women, some of them away from home for the first time in their lives. The night clerk was an important fallback for a student who forgot her key or felt in danger or distress. The job of collecting residence fees was always entrusted to one of the senior staff, because it was a twice-yearly chance to talk personally to each of the young women, to see that all was well with them. If things were not well financially, the staff woman could take the case to the dean, who might then sign a postponement authorization on the back of the dues notice. The system was an example of the holistic technological and organizational practice that used to prevail even in fairly bureaucratized institutions in the 1960s and 1970s. People improvised according to local and immediate circumstances. It wasn't perfect, any more so than the typical computer system, but it was an attempt to work things out, for better or worse, in the closely knit oral culture of the living community.

The dean of women argued her case. The computer's system for running the residence would be blind to all these personal complexities and sensitivities. She was offered an early retirement—whether strictly related to this incident or not, she wasn't sure. The computerization plans went forward as designed.

The computer's power to transform the context and culture of work—what we call the ecology of work—is perhaps the most profound force shaping society in the late twentieth century. Yet it is hardly ever acknowledged, and the voices challenging it are seldom if ever heard outside of a small circle of friends.

Before computerization transformed them, workplaces operated more or less on their own logic, the logic of the particular place, the particular people working, the particular tasks. They weren't the clone of a universal technologic with a monopoly on workplace knowledge and operating procedures. Telephone operators at the switchboard in Midland, Ontario, and Ste Agathe, Quebec, helped take the load off each others' shoulders when one of them would get repeated calls from faithful customers who wanted information, or who simply wanted to chat. Then Bell automated long-distance switching, centralized operator functions into a few remote-control call-centres, and closed these local switchboards.

A chronic-care hospital in southern Ontario let some doctors who were beyond retirement age continue relationships with some of their oldest patients, because the medical director judged that to suddenly switch doctors might hasten the death of a patient. Then a chief executive officer was hired, superior to the director of medicine, and terminated this practice as a risky deviation from proper operating norms.

One by one, various sites in our working and living environment are being clear-cut by the standardizing logic of computer systems. Once you are enclosed within those systems, you have little choice but to adjust to their built-in biases. The old context is gone. The continuity, and related priorities, are obliterated. The old ways and wherefores of doing things have no status except in the tacit knowledge of the older workers and their memories of the good old days.

Job Title: Traffic Operator Position

The revised version of the telephone operator's job prophetically symbolizes the new work model, because the job title coined to accompany Bell Canada's integration of call-processing services in the late 1970s doesn't even suggest a person. It describes a technical function in which human agency, and personality, have largely disappeared.

Telephone operators were finally fully transformed from being active agents of communication, the link between the early tools of telephony and people wanting to communicate, to computer-controlled McJobbers, closely monitored in their performance. The computerized system encloses operators inside their headsets in isolated little work stations and in a wholly computer-defined meaning of work. And it transforms them into a functioning *part* of the telecommunications system—not as an agent of technology, but as a servo-mechanism.

Under the traffic operator position system (TOPS), calls requiring operator assistance are funnelled through a large regional computer and parcelled out to any terminal anywhere across the system that is not engaged in another call. There is no personal preference involved and no continuity among a local or specialized set of callers. It's all technical, like distributed data-processing. The pertinent caller information flashes automatically onto the screen in front of the operator in her (or his) cubicle, and the caller's voice—from perhaps hundreds or thousands of miles away—arrives in the headset of the operator, who speaks the standard phrase, "Operator, may I help you?" The operator's only other function is to enter the billing number or other data into the keyboard and press the "enter" key to clear.

As one operator describes it, "The calls come in through my ears and go out through my fingers, and never once enter my brain." Another describes herself simply as "a robot."

Cut off from any relationship to the now remote and anonymous customers, telephone operators—who are increasingly employed part-time—are also cut off from each other. Before they operated the switchboard as a team, balancing the load of incoming calls up and down the cordboard, but now they are alone with their computer terminal, which strictly controls how much of themselves they can bring to the job at hand. Cut off even from themselves as whole, sentient beings, they start relating to the system as virtual automatons. Their identities start to revolve around the system's feedback on their performance in the only way that counts within the system: their actual working time or AWT, computed as so many seconds per call.

Now this model is spreading into the proliferating range of digitized work such as hotel and airline reservations, banking, insurance, and registration for courses, government, and health services—all delivered through variations on telework. The stepping stones are in place in just about every line of work in every sector of the economy. Once work has been computerized, it can also be readily and almost costlessly relocated to computers almost anywhere: wherever there is a plug-compatible machine and a plug-compatible labour force schooled to accept marginalized bit positions, at lower wages, with few if any benefits and no job security or opportunity for personal development.

Call-Centres and Telework

New Brunswick Premier Frank McKenna has made call-centres and the information highway the centrepiece of his industrial strategy. Through a strategic partnership between Northern Telecom and New Brunswick Telephone Company, the McKenna government has developed a high-capacity integrated digital networking system that has put the province in the forefront of telemarketing and other forms of telework. With this structure in place, plus generous government subsidies, training guarantees, and brochures advertising "no payroll taxes and low workers' compensation rates," plus "labour costs . . . 15-20 per cent lower than in most other Canadian locations," and a literate bilingual population, government representatives have actively solicited work from across North America. They haven't enticed companies to come as corporate wholes with head and hand work combined. They are settling instead for computerized support work—clerical, sales, and service—only.

When it moved a much-publicized 850 jobs to New Brunswick in January 1995, United Parcel Service (UPS) joined three other courier services (Purolator, Federal Express, and Canada Post) with call-centres there, boosting the total number of call-centres in the province to twenty. In a speech to a glossy national conference called "Telework '94" in Toronto, New Brunswick's assistant deputy minister, Information Highway Secretariat, Jerri Fowler boasted that New Brunswick was the "call-centre of North America." By contrast Ontario was keeping fairly quiet about its over two thousand call-centres, including the continuing presence of UPS, which employs thirteen hundred people in Ontario alone.[3]

Statistics Canada put the number of the telework labour force at three hundred thousand in 1995, although the job defies easy categorization.[4] Many tallies include full-time employees who can enjoy the choice of sometimes "telecommuting" from home, plus self-employed professionals who can charge for overhead and can leverage a professional fee structure. But judging by the growth and range of telework in the early 1990s, it could be the largest source of new job growth in this decade.

So far telework includes a lot of customer service work: for instance, CIBC (auto insurance) and Consumers Gas in Ontario; and courier companies, an electronics firm called Camco, plus banks, insurance companies, and other financial-service firms in New Brunswick. But sales is a growing new feature, including airline and hotel reservations and, in the United States, an expanding contingent of "sales associates" for the JCPenney retail chain, working both in sales call-centres and as "at home associates."[5] Look at all the 1-800 numbers popping up these days. Chances are good that there's a teleworker in a call-centre or at home on the other end of the line.

Purolator's one national 1-800 number handles pick-ups, package tracing, supplies ordering, general inquiries about services and customer billing, on a largely cybernetic and self-serve basis. For example, when customers call to trace a package they hear a computer voice prompt and are told to enter the ten-digit waybill number on their phone keypad. A computer-generated voice reports on when the package was delivered. (This information derives from hand-held instruments operated by delivery personnel. The company's trucks are tracked under the ubiquitous gaze of geopositional satellites, part of the company's Satellite Tracking and Reporting System, which collects delivery details and knows to within three hundred metres the exact location of every Purolator truck in its fleet.)[6] Only when the system can't serve (or service) a customer automatically is the call transferred to "a live agent."

For the appliance manufacturer Camco, 51 per cent owned by General Electric, a call-centre in New Brunswick is the logical complement to the company's quick-response manufacturing strategy, through which 50 per cent of its products are now shipped directly to customers' homes. The 1-800 line integrates basic customer services such as order-processing and telemarketing with value-added warranty, repair, and ongoing technical-support services associated with the new product-service concept and related "product loyalty."[7] Furthermore, data analysis from the centralized national service lets the company standardize and better measure customer service and do jobless market research.

Two banks, the Royal and Canada Trust, are using "virtual bank" call-centres as stepping stones to a two-tiered banking system: with a core of bricks and mortar offices with staff providing value-added and possibly pay-per services, and a host of stand-alone kiosks supplemented by do-it-yourself banking from people's homes and cars. Canada Trust's EasyLine service, operated from call-centres in New Brunswick and London, Ontario, offer bill-paying, credit-card, and bank-account updates, plus mortgage application, certain loans and RRSP contributions, and certain related investment information and transactions. It's also a "borderless" service, because the enabling technology was designed for easy North American and even global flows of data, whether it is money, software, or surveillance images.

A "facilities management" call-centre, in Dieppe, N.B., monitors residential, business, and industrial properties "as far away as Australia," according to New Brunswick's Jerri Fowler.[8]

The 1991 census gave a hint of the potential growth in home-based telework. Since 1981 there had been a 40 per cent increase in people working out of the home, with large clusters of people engaged in administrative, managerial, and related work, at 138,000, and in clerical and related work, at 126,000. For women, clerical work was the largest single employment category, accounting for over 110,000 women, followed by service work, employing 98,000 home-based workers.

The census figures didn't correlate income with occupation, but it was telling that nearly 70 per cent of women reported making less than $20,000 a year—and 145,000 (close to 30 per cent) reported earnings of under $10,000. In one of the few studies on homeworkers, a masters thesis survey of home-based clerical workers found incomes ranging from $300 to $14,000, with a mean of around $7,000.[9]

Meanwhile, companies involved in expanding and running the infrastructure are making money by selling the new technology required to make telework happen. Rogers Communications is promoting a CableLink-Work service for home-based workers. The service offers multimedia online applications and file-sharing at speeds that it boasts are a thousand times faster than phone-modems can deliver. Northern Telecom offers a "visual interactive technology" system for teleworkers and telecommuters.

The federal government has actively encouraged teleworking in the public service, with one department spending $6.5 million on lap-top computers in one year-end buying spree alone.[10] Furthermore, "By committing to use telecommuting themselves," the general manager of Stentor Resource Centre told a researcher, "governments legitimize it in the eyes of other employers, and more importantly . . . create a stable market for telecommuting products and services."[11]

The phone, cable, and equipment suppliers make money selling technology, and they make more money ferrying information back and forth around the world, keeping their communications lines busy. They have the same vested interests in expanding the infrastructure and keeping it operating at full capacity as truckers, truck manufacturers, and gas companies a generation ago did moving goods around on superhighway systems, and a generation before that, other corporations did moving goods and people around by rail. And their interests neatly dovetail with those of corporations working the information infrastructure for the lowest wages and the highest payback rates, and striving always to balance workloads across vast interconnected systems to, in turn, operate their various facilities at full capacity.

Computer-Monitoring and Performance Measurement

Computer-monitoring and performance review are a sign that computerization has reached the stage at which work can be completely defined by and contained in computer systems. At that point its performance can be removed from the direct supervision of an actual corporate structure and relocated to any site where there's a plug-compatible system, because the corporation's supervisory gaze comes built in with it. If the job can be done remotely, its performance can be monitored and managed remotely too, because it can be measured in strictly computer terms. As this happens, computer-monitoring performs a second, more subtle task. It seals people into a completely computer-defined sense of their work, which they are then almost forced to internalize.

Computer-monitoring also represents a realization of the superpanopticon that Michel Foucault described in relation to data bases, the panopticon being a prison structured so that inmates could be seen by the warders at all times. Now the all-seeing eye of the central computer focuses in on the microlevel of personal experience and individual jobs.

Equally, monitoring—when coupled with "cultural training"—could fulfil the original techno-utopian hopes associated with the panopticon. The dream was that prisoners would willingly come to comply with the prison rules and regulations and no longer require constant watching. Once all other ways of being and being seen to be were cut off, they would identify with Big Brother, as it were.

Computer-monitoring exists, explicitly or implicitly, in almost every worksite today. However, its use follows clear hierarchical and gendered lines: production and support-staff workers are monitored more than managers and professionals. Women are monitored more, and more thoroughly, than men.

In factories sensors embedded in the product being made keep track of exactly where it is along the production-line and indirectly report on how fast each worker involved is working. In one factory at least, workers also have to "badge" into and out of the lunch room, using the magnetic stripe on their identification badges.[12] Geopositional satellite systems, with beacons embedded in cars and trucks, allow courier and taxi companies to know exactly where all their drivers are at any particular moment. Nurses, food-service workers, and many other health-care workers in hospitals have found themselves tied to increasingly systems-defined job descriptions through strict and detailed accounting and reporting requirements associated with "total quality management."

As Dave Robertson of the CAW put it, "What electronics does is make the work process visible to management and therefore controllable by management."[13] The software built into the telecommunications systems in New Brunswick call-centres includes monitoring and performance evaluation. The software also analyses call data to detect emerging "routine call" patterns that can potentially be automated through more computer-voice programs, and it provides ongoing feedback for "scheduling agents properly for peaks and valleys in call activity."[14]

Computer-monitoring has been used to school workers in a more technical, systems-defined conception of the work they are to do. In 1985 fully 59 per cent of CUPE locals reported that computer-monitoring "had resulted in work and productivity being scrutinized in more detail." Furthermore, "The overwhelming use made of monitoring was to discipline workers. Monitoring results were also frequently used to provide workers with feedback on job performance and to make judgments about whether or not workers' employment should be continued or discontinued."[15] Some older women working as supermarket cashiers in British Columbia wrote a letter of protest after they had been called into the office by management because they were "supposedly having problems on the scanners and were below average in dollars per hour."

According to the letter, "The whole staff knew why we were singled out and kept there for nearly two hours. We were never so humiliated in our lives."[16]

In the letter the cashiers argued that comparing cash on an hourly basis was meaningless. "What is average? The elderly customers with their small orders, and their pushcarts wait in our lines because they know that we'll pack their groceries properly, and treat them courteously, even while they count out their change to pay. . . . Customers know us and bring their grocery problems to us to verify."

Judging cashier performance on the basis of average cash flow per hour or shift is indeed meaningless—that is, when the role is defined within the context of the human environment, where grocery stores are valued as social institutions within geographic (and cultural) communities. However, strictly data-performance measures do become meaningful when the context has been transformed into an electronic utility for expediting the flow of food purchases in the fastest and cheapest way possible. Furthermore, unless some outside factor intervenes to prevent it, the technical system will tend to transform the cashier, willingly or not, from an agent in a relationship with a customer into nothing but a set of remote-control fingers, disembodied digits acting as an extension of a price-scanning and payment-processing system. It will also transform customers into passive objects, scanned and computed just like the merchandise and almost as interchangeable as the part-time staff who putatively "serve" them. At that point, why not "opt" for remote "teleshopping," especially as the gap between the haves and have-nots heats up with muggings and angry demonstrations, making the commons of public spaces less and less hospitable?

The use of computer-based performance information as a central criterion of performance is transforming the workplace. A comparison of monitored and unmonitored clerical workers who were engaged in processing claims in a large Canadian insurance company revealed that the unmonitored employees tended to be unaware of the production quotas that management had set for their work. But the monitored employees were not only aware of the quotas; they regularly talked about their work in productivity-related terms. "People are always comparing themselves to each other on production," one respondent said.[17] Furthermore, the clerks had become so oriented to "the count," as it's called, that they became accomplices in redesigning their jobs from a collegial mix of teamwork and individual effort into one-to-one relationships with computer terminals and a narrowly quantifiable range of tasks. Previously a difficult claim had made a day's work more interesting, but now it had become an imposition dragging down the clerk's daily count of claims processed. As one respondent said, "The computer can't tell how hard the claim was, so I end up looking bad." Another asked, "Do you want ten phone calls done correctly, or twenty-five just pushed through?"[18]

The report added, "This same aversion to difficult claims surfaces in complaints of eroded teamwork and cooperation among processors within a unit." Instead of co-operating, the workers competed in a struggle to process the easiest, most straightforward of the claims. As this and another study by Patricia McDermott documented, the productivity measure and related monitoring imposed an almost adversarial work culture, and a new ecology of work in which people related to the machines and the operating system rather than to each other.[19] Indeed, some keener workers come to like computer-monitoring, because it forms an objective wedge between themselves and others who might be "sloughing off."

Another study revealed how a Canadian public-sector company, Air Canada, had been transformed into something like a production-line operation through computer-monitoring. Before computer-monitoring and performance measures were introduced in the mid-1980s, customer-service agents at Air Canada had a basic latitude in dealing with customers, helping them find the best routes at the best price. Under the new system their work was redefined around a computer-monitorable standard called "telephone service factor" (TSF), which measures whether an office responds to 80 per cent of calls within twenty seconds. Another computer measurement, "in-line time," calculates individual productivity.

The TSF puts staff under permanent productivity stress. As one of the managers reported, "We staff to 70 per cent. We can't afford to staff to 80 per cent."[20]

The "in-line time" standard adds a different kind of stress, because it only counts what management defines as value-added time. Time spent finishing a file after a call has ended, calling a customer back, and simply giving fare quotes is not counted. Value-added has been redefined more narrowly to mean selling more product. A new management motivation program defines "quality of effort" in terms of converting shoppers into passengers. In one office weekly performance results were posted on a notice titled "Let's team up for quality and sales." One week the performance, measured as the percentage of potential bookings converted into actual bookings, was 32 per cent. The notice then posted a target of 45 per cent for the following week.

Agents' work stations became "Reservation Selling Stations." One agent said, "This is fast food in the airline industry. The seats are our hamburgers." Another agent called it an assembly line, with workers reduced to robots. In the film *Working Lean*, Air Canada agent Karen Millington said, "All the user-friendly words they're using now are really hiding a really dark, menacing face."

Some staff work through their lunch hour, saying that is the only way to meet the performance standards. "They want us to compete with computers," one said. "You program yourself," another said. "This was a nice place to work two and half years ago, but I've seen it go downhill continuously. They're going to turn it into another prison like Canada Post."

Another agent said office attitudes were changing. People who once were "nice" were becoming not so nice. "People yell in the office quite often. Cathartic releases are common around here."

Four out of five agents said that their jobs had become regimented and routine. Most of them saw no opportunity for advancement or personal development. Muscle strain, eye strain, and headaches were common, and some workers had developed skin rashes. Two out of three reported stress, and over two-thirds linked stress to computer-monitoring.

Management consultants are trying to mitigate the stress by making the systems more reasonable, enlisting workers as willing participants in computerized work with computer-monitoring and performance measures—and, increasingly, related payment schemes. The Computer and Business Equipment Manufacturers' Association advises clients to take time to explain to workers "why and when their work is being monitored." Companies should "reward individuals appropriately" and not "continually drive up production standards."

A study by the University of Western Ontario's School of Business Administration adopted a thermostat feedback-control system as its model. That system, the researchers believed, would provide a method of predicting which work "conditions or environments" were most likely to produce the desired results. The research isolated factors that might contribute to workers' acceptance of performance monitoring and would help them to internalize its standards: factors such as defining the work completely within the terms of the technical operating system, so that measuring performance in the purely technical terms available through computer-monitoring would seem reasonable and meaningful to the employee. True enough, the study found that the clerical workers willingly co-operated in excluding from their work those activities that would handicap a high count. At the same time, though, the researchers also found a need for a complement of what they called "high touch" procedures in the form of good human-relations management. This would ensure that workers accepted computer-monitoring positively, as an unbiased source of performance review and self-motivation.[21] This is where co-management "teams," "total quality management," and other "worker-empowerment," worker-involvement programs come in.

Training for Compliance

A decade ago I was asking: training for what? All my research indicated that more skill, knowledge, and creative involvement were being sucked out of jobs than put into them. Certainly there has been an increased demand for professionals, particularly those able to work in a fully networked computer environment, and for people to build, upgrade, and manage such workplaces. But the biggest trends have been towards jobless economic growth, and the simplification of work into McJobs done by an ever-increasing contingent labour force. Other research documented a huge disparity between the level of computer knowledge and skill in Canadian society and the job opportunities available for applying and expressing this knowledge and skill. One study found that 80 per cent of high-school and university graduates had taken some computer training; but of the 50 per cent of these who were working with computers on the job, the vast majority were restricted to simple data-entry, data-processing, and word-processing.[22]

This doesn't mean there's no need for training, but it puts the scope and scale of the need into perspective. So does a 1994 survey of seventy thousand adults using Toronto's Daily Bread Food Bank. Three-quarters of them had lost jobs since 1990. Their job skills? As *Globe and Mail* columnist Michael Valpy reported, "Those out of work are more likely to be computer-skilled than labourers, more likely to be teachers than taxi drivers." The largest single category of food-bank users in the survey were office workers, at 19 per cent, followed by construction workers, at 17 per cent, then professionals, at 14 per cent.[23]

Part of the puzzle is explained by the fact that training and retraining, "individualized" as computer-based learning, have become big business.

Some of the biggest companies associated with the information highway and globalization generally are promoting privatized training. In Ottawa, the Corporate Information Technology Institute charges $12,600 for an eleven-month combination of training and work experience in the high-tech sector. Training has become a make-work project for enterprising private corporations, with the unemployed and underemployed representing almost a captive market for their courseware and their questionable pitch, namely that the equation of "more education equals a good full-time job" still works—or at least it might help you get a foot in the door.

The other part of the puzzle is explained by the content of much of the training. A good deal of corporate training is training for compliance. It's not training geared to personal growth and development, but training geared to binding workers to the corporate mission, pulling them into their dynamo of faster, cheaper equals better.

Writer Jamie Swift calls this dynamic "attitude-adjustment-as-training."[24] In a sense, it can be seen as a species of advertising, with its language that shapes not just our appetites and attitudes but even our identities.

Essentially, advertising comes down to rhetoric: words and images removed from their original context of meaning and manipulated simply as "signs" to stimulate and express desire. Earlier critics emphasized the external manipulation of the passive viewer, as an object being worked over by these signs and symbols. Others have stressed advertising's subliminal effects, which work on our subconscious. Postmodernists such as Jean Baudrillard go beyond this. They argue that viewers aren't passive objects swept along by advertising's messages. Rather, viewers participate in the "invented" realities of advertising. Advertising, Baudrillard argues, is a language that viewers speak along with the advertisers.[25] Like so many postmodernists, he ignores the hard political questions around agency and authorial voice in that language. In other words, he ignores the gagging of the original voice of the subject, while the (produced, paid) voice of the advertiser takes over. But Baudrillard's point about active involvement is an important one. In other words, it doesn't matter whether the viewer believes an ad for floor wax in which Prince Charming magically appears in the kitchen of the harried housewife offering X brand to her like a rose. Instead, the ad sets up a new representational logic (or illogic) in which floor wax can be romance. The viewing public buys into that representation and, hence, into consuming the product.

Essentially then, ads offer us an image of ourselves, and we consume this image. This is a simulated, invented identity, but one that another theorist, Roland Barthes, maintains is becoming the new culture, operating at the level of coded signs associated with fashion, a language we speak as we keep up with fashion. This analysis seriously underrates the durability of traditional grounded cultures. It also discounts people's capacity to retain their own identities and thus their ability to be social agents, modifying the technological landscape even as it modifies them. Nevertheless, its pervasiveness, and usefulness to business, need to be taken seriously, including in cultural "training" programs.

The corporate mission statements and productivity slogans function like ads: they are an invented code or language, and they speak the participants into existence as those participants consume the cultural training and take part in "Excellence" teams.

The words such as "empowerment" and "co-management" are, like key words in advertising, divorced from any real context and re-engineered in the corporate-communications script to foster a new attitude and identity.

As identities associated with traditional jobs and social relations with co-workers are drained away, a new identity is offered. Workers are invited to consume the corporate image. They're invited not only to rise above class and gender exclusions, but also to identify with the corporate mission and become virtual corporate agents serving customers with quality products around the world. Rick Dove, one of the principal associates with the Agility Forum and chairman of his own consulting firm, actively promotes "shared-vision workshops" as a cultural counterpart to what he calls the "Agile Journey." The purpose of these workshops, he says, is to cultivate "evangelical support for the common vision."[26]

Summarizing the role of training during the late 1980s and early 1990s, a human resources manager at a large electronics firm says, "High-impact communications programs clarified the (corporate) vision, and training programs encouraged individuals to explore the vision and become part of it."[27]

An employee-orientation manual introduces the company's "excellence" program as "a new culture . . . a new way of conducting business." Elsewhere, the program is described as a way in which workers learn to "see through the eyes of managers."

Coupled with such sophisticated orientation programs, the logic of measuring and monitoring systems threatens to fulfil two important philosophies of control. One is Taylorism, and the other is associated with Jeremy Bentham's original vision of a prison panopticon.

Taylorism and the Panopticon

Management and management literature in the late 1980s made a great show of moving beyond Taylorism. They vilified it as a big, bad bullying external form of control. They repudiated it as a waste of productivity through alienated workers, and a nineteenth-century anachronism. Meanwhile, the new management style of employee involvement went to work fulfilling Frederick Taylor's original idea, that "joint obedience to fact and laws" would replace "obedience to personal authority."[28] What this meant, in other words, was not external control, but internalized control by a compliant workforce. The method, which came to be known as "Taylorism," was for management to fragment production into a set of prescriptive tasks, to study how best to do each of these tasks, and then to supervise their execution based on scientific-management principles. The original method was crude: time-motion studies conducted by industrial engineers hired as consultants. More recently the method has been refined by enlisting the assistance of workers in place of time-motion consultants, in a joint pursuit of "facts" such as productivity standards and "laws" such as competitiveness. The contradiction in all of this—namely that management still controls the definition of these facts and laws—is buried under the corporate-cultural rhetoric of post-Taylorist employee empowerment. In fact, however, it is a recycling of Taylorism as "social Taylorism" at the microlevel of the workplace.

Jeremy Bentham's prison panopticon rested on a similarly arrogant, and naive, faith in a prescriptive set of mechanistic norms to which everyone in society could be adjusted, even willingly. To Bentham, having a set of watchtowers constructed around a prison so that all prisoners could be kept under surveillance at all times, with minute records kept on their actions, was not a sinister act of oppressive control. It was a device for encouraging compliance and conformity. The idea was to "deflect the criminal's mind from the irrationality of transgression to the rationality of the norm. . . . With no escape or reprieve from the Panoptical eye, the prisoner would accept the authority of the norm with its rational system of pleasures and pains."[29]

There is little evidence that total-quality management programs pay off in actually reducing costs and increasing productivity, and little research available on the subject. But perhaps productivity has little to do with the primary goal. As an office worker at the CAMI automotive plant in Ingersoll told me: "Nobody knows how much money is spent on these programs. Nobody can put a number on them—

and this is a company that counts friggin' paperclips!

"The training, the uniforms, the exercise program, the quality circles (done on people's lunch hour), they all work to create this one effect. They take away your individuality. They make you conform, yet nothing is ever said. It's insidious."

For example, she said, the women no longer take their purses with them when they go to the washroom. They don't brush their hair, or put on lipstick or any other makeup while at work. "We have a culture of people here now who've altered their whole behaviour without a word being said to them. And they don't even realize it."

With more money being poured into cultural training, and into making computer-monitoring more socially acceptable, the socializing effects of these programs become significant. If computer-monitoring is being modelled on thermostats—with their controlling units programmed to respond automatically—the goal of fine-tuning them might go far beyond muting resistance and preventing sabotage. It might be to re-engineer workers' consciousness so they react unthinkingly to prompts within the system of which they are a part.

In other words, the goal might be getting people to "turn turtle," in McLuhan's sense of the word. People begin to think as they've been programmed to think; harmonized with the corporate mission statement, synchronized to the drumbeat of computer logic. They become, rather strangely, both model prisoners in a panoptical silicon cell block, and model turnkeys. In submitting to the computer as a meaningful source of performance feedback, workers can no longer use their own intelligence, intuition, or judgement. They're almost forced to let the system's quota standards—for instance, the actual working time (AWT) for telephone operators—carry them through the day.

In Québécois filmmaker Sophie Bissonette's film *Quel Numéro, What Number?* a group of TOPS operators confess to becoming, almost unconsciously, addicted to checking their AWT. "The only thing left is the machine, and the machine only tells you one thing: your AWT," one of the women explained. "Soon you're hooked; you try to reduce your AWT; you try to beat the machine."

Whether because of this tendency or not, in spring 1987 some Bell operators in Ontario suddenly found notices posted in their TOPS cubicles. Part of a new campaign called "Superservice," the notice urged operators to "Go ahead. Get personal. Make their day!!!!" and listed a number of phrases that they could use to do this: for example, "Thank you for waiting" and "Have a nice day."

At least one of the operators I talked to coldly ignored the poster. "They want me to do my work in so many seconds? I don't have time for all this other nonsense," she told me crisply. Shrugging, she added, "I'm pleasant. That's all I have to be."

In the Quality Electronics study, a machine operator reported so much pressure to meet production targets that if someone left the line, others would "yell at them."[30] He also reported people tattling on each other, and of shifts competing with each other: "If the day shift ran 1500, then the afternoon shift comes in and says, 'Well, if they ran 1500, we've got to run 1700.'"

A woman with ten years' seniority finds that the "team concept" means "a lot of people running to the boss and saying, 'I can work harder than this person.' So they're backstabbing their fellow employees. They get disgusted if you can't keep up with their pace: 'You're lazy, they're not.'"[31]

In workplace after workplace, under the rhetoric of "quality" and "excellence," the cultural environment is being degraded. Hospital workers are squeezed between hospital cutbacks and new computer-monitored performance standards.

"It's dog eat dog. Everybody is nervous, everybody is tense. . . . Supervisors are even saying that (fellow workers) are going to watch and report back."[32] The report documented supervisors screaming at workers, and workers "yelling more at each other and finding it more difficult not to yell at the patients and their relatives."

Yet this could become the new culture of work, especially in the isolated and isolating atmosphere of call-centres and home-based telework.

Carol Van Helvoort is a teleworker for Pizza Pizza. In a split-shift day that runs from 10 a.m. to 2 p.m. and 4 p.m. to 8 p.m. she processes tele-orders for pizza at a rate of nearly one a minute through a company computer set up in the bedroom of her high-rise apartment on the fringes of Toronto. For this she makes $7 a hour, plus 1 per cent of sales over a certain total. She is a member of a union, one of the few teleworkers with that protection, whereas many women she knows have to rent the computer and modem from the company subcontracting out the work.

"We're monitored constantly, on the phone, on the computer," she says.[33] But mostly it's the boredom and isolation that gets to her: being trapped in a tiny subset of a social transaction, repeating it endlessly minute after minute, hour after hour, shift after shift, year after year: the same standard phrases, the same keystrokes, the same narrow menu of predefined choices both for her and all the customers out there in

Metro Toronto. The isolation is also imposed on her family as the work takes over the living space of her apartment for two chunks of the day. Her son can't have friends in. It's difficult to have visitors because the phone line, which also serves to let people into the foyer downstairs, is totally occupied by the virtual presence of Pizza Pizza, with its non-stop stream of customer orders.

She mentions other women whose younger children have learned to adjust to the imposition. "But my God, they shouldn't have to adjust. What does it do to the family?"

A University of British Columbia study of homework and family life reports, "None of the homeworkers [men or women] cited improved family ties as an advantage." Many had "unresolved conflicts between the responsibilities for their family and their work."[34]

Researcher Margaret Oldfield finds that homework typically results in more work for mother, as it reinforces gender stereotypes about women and the private sphere and the invisibility of work going on there. She also links this to an International Labour Office prediction that the issue of telework will dilute the debate on public child care and to U.S. right-wing politicians' calls for computer loans for stay-at-home mothers.[35]

Carol Van Helvoort also worries about women who are abused. "They don't have the safety of work to go to. It can be a real problem."

Being stuck at home, going nowhere to work, affects her as well. "You lose a lot in conversation. You don't have the small talk. As far as doing your housework, forget it!" She says that before, when she worked out of Pizza Pizza's office, her dishes were always done before she left for work. "Now, I don't bother. I used to curl my hair every day. Now, forget it! I don't bother to do my nails. You lose the momentum to do things. You stagnate. You don't even bother getting dressed half the time. . . .

"And you smoke more. It's the boredom. You can't move!"

She would quit tomorrow, she says, if she had the choice. "But with the market the way it is today, forget it. There's nothing out there. If I want to work, I have to pay the price."

This could be the brave new world of work for a lot of people. Its implications go beyond the "divisible," of what's happening to others, into the "indivisible," of what's happening to us all, because it involves a profound shift in the ecology of work and with it the culture of everyday human interaction. Our social and cultural environment can only withstand so much erosion and degradation before we lose the capacity to sustain it as a healthy, inclusive whole. Social divisions will rigidify. Cybernetic apartheid and digital alienation could become entrenched.

Repetitive Strain Injury and Layoff Survivor Syndrome

Individually, the strains of working in isolated job cells tied to a computer terminal are dramatized in what's emerging as the new economy's prime occupational hazard: repetitive strain injury (RSI). The U.S. Occupational Safety and Health Administration has called it "the occupational disease of the 1990s." Others call it an "epidemic."[36]

More than 50 per cent of the workforce is at risk from RSI, particularly those whose work involves "rapid, repetitive motion and awkward posture, such as cashiers, assembly-line workers, bank tellers, airline reservation personnel, telephone operators, data processors, mail coders."[37] Women in particular are at risk, because they're concentrated in jobs that tie them tightly to computers doing simplified, repetitive, boring work, although young men are joining the risk circle too.

The study of women working the production line in the case of Quality Electronics noted that the work area "was notorious for RSI; at one time, 30 per cent of the workers had RSI."

A film on lean production in the auto industry showed many workers with tensor bandages on their arms. "You become almost robotic," one of the workers said. "I don't think man was designed to do the same thing day in and day out."[38] At the Ingersoll CAMI automotive plant there are a lot of bandages, especially along the final assembly line, where the work is the most taxing and pressured. "We have some real strong body-builder types in here," one of the union stewards says. "But their muscles are breaking down."[39]

RSI is still not recognized by all workers' compensation boards—with British Columbia's board reportedly denying many cases.[40] As well, there are concerns that part of the deal made between UPS and the New Brunswick government was that compensation regulations wouldn't cover claims for repetitive strain injury or stress in call-centres.[41]

Yet teleworkers could be the most vulnerable of all to RSI—not only because of the narrow, repetitive, and completely computerized nature of their work, but also because "psychosocial factors of the work environment also appear to play a role in the development of RSI." These factors include control over the work (or lack thereof), quotas or productivity bonuses and other work-pressure factors, plus opportunities for friendship (or lack thereof).[42]

In the United States, workers' compensation claims related to stress tripled between 1980 and 1990, while a major study on stress isolated lack of control on the job as the biggest factor responsible. "The primary risk factor [for heart disease] appears to be lack of control over how one meets the job's demands and how one uses one's skills," the authors reported.

> It is not the demands of work per se but the organizational structure of work that plays the most consistent role in the development of stress-related illness. . . . It is not the bosses but the bossed who suffer most from job stress. The most common problem is stress among low-status workers who bear equally heavy psychological demands but lack the freedom to make decisions about how to do the work. . . . This stressful job fits, embarrassingly closely, the specific job design goals of Taylor's scientific management.[43]

Everywhere people are being forced to work past their levels of tolerance and personal equilibrium, to do twice the work in half the time, to compensate for less-skilled, part-time staff replacing competent full-time staff, to fill in the gaps created by sickness, attrition, and workplace disabilities like RSI.

At best, stress-management seminars merely mitigate the problems, teaching tricks for coping with stress, not preventing or stopping it. Cynically, companies, such as MEC Air Technologies in Toronto, are pumping aromas into workplace air systems (not always with workers' consent or even knowledge), using peppermint to massage workers into alertness or jasmine and lavender to massage them into a false sense of relaxation. Research behind this technology is geared to promoting human performance rather than human equilibrium and well-being.[44]

The stress associated with "layoff survivor syndrome" combines guilt at still having a job when so many have lost theirs and anxiety at the thought of being the next in line. The syndrome leads to depression, frustration, and anger. It also results in lower productivity and performance as people postpone making decisions that might bring them flak and spend enormous amounts of time chasing the latest rumours about layoffs.

Whose brave new world is this? A world fit for cybernetic dynamos and the techno-cowboys who think they're staying on top of them, perhaps; but not a world for most people.

So, what can we do about it?

PART 3

RESPONSE

Restructuring for People

6

A Plea for Time — A Plea for Our Times

Let the meaning choose the word.
– George Orwell

Propaganda begins where dialogue ends.
– Jacques Ellul

So many people have a limited view of reality, not being able to feel the reality of others.
– Margaret Laurence

THE OFFICIAL DISCOURSE on technological restructuring won't help us come to terms with what is happening, or help us to deal with it as the urgent social issue it is. It won't help us because it's centred not on the needs and priorities of most people, but on the priorities of the global corporate economy. It is framed around that economy's need to globalize and downsize, to create a global labour pool with minimal social safety nets, and to do whatever else is required to operate efficiently and expand its marketing space.

People enter into the discourse only in terms of that economy—as new skill-sets needed, or as redundancies to be adjusted into retraining or workfare programs on the margins. Furthermore, the discussion has focused almost exclusively on the state as the agent to manage the social-adjustment aspects of the restructuring agenda and to mitigate any untoward effects. Government bureaucrats, in consultation with experts from business and labour, are qualified to script what's to be done, and to do it. The rest of us are bystanders.

This must change on both counts.

A new critical discourse is required to shake the official adjustment agenda's sense of glib inevitability, to take charge of the restructuring process, and to reassert meaningful public control over the key technologies so people everywhere can start programming them to serve the needs of people and human communities. This new critical discourse must be centred in people, in the social context in which restructuring is occurring, and we must all participate if it is to be effective.

Things aren't going to get better simply by having governments felicitously ease up on the social-spending cuts before the next election. Nor will they improve when the economic indicators clear us from the next recessionary downturn. Things will get better only when we reclaim the economy for people and the common good, when we restructure and reprogram its infrastructures to involve people and extend the scope of the human or social economy, and when we renew the necessary public regulatory tools for holding the corporate systems economy accountable for the effects of its infrastructures on the social environment.

The new critical discourse must be centred on people as citizens, not as corporate and corporatist groups. It must be a discourse of embodied stories and analysis, naming and comprehending the human experience behind the silicon-smooth mask of "restructuring" portrayed in the official discourse. People everywhere must speak out, must attest to the lived reality behind "downsizing," "contingent" job status, teleworking tasks, and computer-monitoring/performance measures. If the global village is not a place for people, only for servo-robots, we need to rebuild the groundwork of the information highway, using our money as taxpayers, consumers, entrepreneurs, and the new self-employed—or have nothing to do with it.

The new critical discourse must also be a discourse of action: practical policies for taking control of the restructuring agenda locally and more broadly through the information highway. We must reconstitute ourselves as a civil society: no longer standing by passively on the receiving end of policies and opinions delivered by the experts and various other technocratic authority figures, but as participants in the social environment.

The Case of the Midland Operators

It's worth revisiting the story of the Midland telephone operators to understand *how* this can be done, and to realize that it can be done.[1] The telephone operators in Midland, Ontario, didn't resist corporate restructuring as such. When Bell Canada announced that it was closing the Midland telephone exchange in January 1984, the employees there simply saw this as impossible. They had nothing against the automated call-switching system (TOPS) that would centralize all Ontario long-distance calls outside Toronto into a massively efficient integrated data-processing system and make local switchboards redundant. Nor was their quarrel with closing local exchanges as such, or even having to commute to work at the TOPS call-centre in Orillia.

They just kept thinking that Bell simply didn't get it. Bell management didn't understand the full meaning of telephone service. Certainly the new system didn't compute with the type of service they had routinely provided out of the Midland telephone exchange—some of them for thirty years and more. While, for instance, some of them had carefully taken on calls from the old lady who had forgotten that the hair dresser was just around the corner from the IGA, or from the cottager who needed boat repairs, or from someone who'd forgotten the closing time at the Bank of Montreal, other women along the switchboard fielded the rest of the incoming calls quickly enough so that they could all still run the exchange at the best call-processing rate in the region. It was a community information service housed in a building not far from downtown, and it dated back to 1886 when a Miss Lillian White ran Midland's first telephone exchange out of her father's grocery and confectionary store on King street. You couldn't just shut off that service line like a tap.

Yet that's what they worried might happen. As they learned about the centralized dispatching software driving the new TOPS system, they sensed that the system would cut them off from the local telephone users, and the local community would be cut off from them. It didn't make sense, and that's why they resisted the company's restructuring plan.

With the help of the local and national office of the Communications Workers Union, they organized a public-information campaign and drafted a petition calling for a public hearing to negotiate and discuss the issue. With 10 per cent of the town's entire population (of fifteen thousand) signed up, they sent their petition to the federal minister of communications, who forwarded it to the Canadian Radio-television and Telecommunications Commission (CRTC), where it was forwarded to Bell for comment. Anticipating the hearing, they drafted their own plan to modify the TOPS conversion plan. Their idea was to computerize

Midland, but in a way that would preserve local operational control so they could still run it as a community information and referral service—a bit like the community freenets of the 1990s.

Several months later the women received a careful and courteous letter from the CRTC's director-general for regulatory matters, explaining that the commission had certain "judgemental thresholds" for determining whether a public inquiry was necessary, all of them involving technical standards of access to and treatment within the telephone service as system. The Midland situation did not fit any of these criteria, so a hearing was not in order.

In fact, the CRTC's job is not to mediate between different visions of telephone service—for instance, communications as culture versus communications as a transmission system. Its job is to ensure that the overall operation is run efficiently as an information-transmission system. Equity and justice are defined accordingly, in strictly technical-systems terms. Adjudicating competing claims for designing and running telephone services and negotiating a compromise between different operating philosophies or logics are matters outside its jurisdiction. Those are parliamentary matters. But Parliament refused to act.

The issue did come up in Parliament, at least briefly. After lobbying the opposition parties as well as the Liberal government, the Midland women and their union representatives gained the support of David Orlikow, the New Democratic Party's communications critic of the day. A month before the exchange was closed, Orlikow filed a Standing Order 21 in the House of Commons, calling for legislation "to ensure workers some right of control . . . of new technology" and deploring the "human and economic suffering" resulting from a case like Midland's, where, he said, there was no opportunity for consultation and negotiation. But the government isn't obliged to respond to standing orders, and the media rarely report on them. Nothing came of the intervention, and the closure went ahead on schedule. The local newspaper covered it with a story headlined "The end of an era."

The Midland story contains several lessons. The first is that it *is* possible to collectively resist the restructuring agenda. But resistance doesn't emerge out of nowhere, nor does it involve simply refusing the given agenda. The women of Midland and their union mounted an impressive campaign of resistance to the corporate agenda because they believed in their own sense of what telephone service was all about, and they derived their own idea of how the technology could be organized differently from their own experience. Their approach was grounded in their own technological practice and tacit knowledge as well as a deep understanding of the context (the local community) in which they did their work. The women had always organized the technology of the switchboard to serve both the corporate logic of system efficiency and the community logic—and they did this not only to accommodate customers' various information needs but also this operator's bowling schedule and that one's Heart Fund meetings.

Their sense of what telephone service should be in the future emerged directly from what it had been in the past and what it was in the present. They defined the technology holistically, in the context of the living local community; they didn't define it prescriptively, according to remote systems plans and abstract service indicators. They judged it the same way, and in the same terms: through the voice of experience in the living oral culture of their work and the community they tried to serve.

In other words, they had their own discourse on technology—one based on real life, not on an abstract model of life in which the future can always be fine-tuned to fit certain preplanned priorities. I'm convinced that this is an essential starting place for all of us who want to resist the adjustment-to-restructuring agenda in which you either scramble up the retraining ladder and compete in the brave new world of digitized work or, as one techno-guru put it, "You're toast."

Towards a New Critical Discourse

Resistance starts at the level of perception, in reclaiming the power of naming from the official media and their anointed panoply of experts: all the people speaking for us, speaking us into existence as objects to be adjusted to the global economy, or as deserving or undeserving victims pushed to the margins. Resistance starts when we speak for ourselves, validating the voice of experience and the claims for justice embedded in the context of human experience. A new critical discourse must be centred in the here and now of technological practice. It must articulate the lived context of it, when people are replaced by computers or reduced to working as disembodied hands and voice boxes under the control of computers. The discourse must begin where things are made and services are provided and used. As it does this it will not only break the immobilizing silence on technological restructuring as an urgent social issue, but will also illuminate how choices can be made differently so that people can program and otherwise negotiate their way into meaningful participation in the new economy. And it will affirm the power that people have at their disposal through their tacit knowledge and experience—power that will enable them to leverage the gap between the theory of technology as all-powerful and dependable and its costly fallibility in practice. That gap represents an opportunity for people to negotiate more control and involvement for themselves on their own terms.

But this isn't easy. Margaret Laurence once referred to "a crisis of the imagination" in the discourse on nuclear arms and disarmament. She referred to jargon words like "megadeath" and "surgical strike." She could just as easily have referred to "downsizing" and "restructuring," and to people suddenly becoming "surplus" and "redundant," or to the general spread of Newspeak in our society.

Breaking the silence on felt experience also takes *time*—the scarcest and most spoken-for dimension in the lives of all the people I talked to in preparing this book. Most people are already locked into technological time. This is the artificial time of contracts, work schedules, and child-care closing times, of meetings planned months in advance, each with an overflowing agenda that everyone has to react to in limited time slots—using acronyms and other telegraphic language in the interests of efficiency. People have little or no time left over to dwell on what they and others around them are experiencing, to fully experience that experience.

In an essay called "A Plea for Time," Harold Innis argued that if there was any hope of restoring balance and stability in our sped-up global technological society, it lay in recovering a sense of time.[2] By this he meant natural time associated with the seasons, as well as continuity, conservation, and memory. As a strategy of communication and cultural survival, Innis's plea for time was a call for local and personal communication—that is, for short-distance communication—to offset the more commercially and monopoly-biased media of fast, long-distance communication. He emphasized face-to-face dialogue and the spoken word as the site and the medium for coming back to one's own self, to know one's self on one's own terms—as opposed to the terms given in the official texts and discourse.[3]

Innis's plea has much to teach us in reclaiming our powers of naming and in naming the lived reality of the new economy as we're experiencing it, as subjects and citizens. In the new (postmodern) politics of identity and self-identification, taking our time to perceive the world and our experience of it on our own terms is perhaps the first political step.

The consciousness-raising movement of the 1960s was vital to women's liberation in the regaining of a sense of self and identity that had been so wholly dictated by men and their media that most of us didn't know what it meant to be a woman on our own terms. In consciousness-raising groups we took our time to find our own voices, to speak the truth of our own daily and bodily experiences, or at least we tried. That was the beginning of feminist discourse for my generation of women, anchored in the catch-phrase "the personal is political." Having named ourselves into existence on our own terms, we could then start to act on those terms as well.

The process was flawed, limited, and abused, but it also did a lot of good. I also think it can serve as a model of resistance at the level of perception. It forms a model because of its orientation to time—the organic time associated with lived experience, articulated as stories. It is also a model because of the medium involved: oral communication. Its grammar is fundamentally different from the language of the official discourse, with its planned agendas, its texts and specialized experts, and its remote centres of corporate authority. As Innis understood, the oral culture of conversation at work and at home and of dialogue at conferences, in workshops, and in seminar courses at universities permits a nuanced expression of particular realities in particular contexts. It validates them as it makes them real.

This type of critical discourse has always existed in the oral culture of the union movement, although for much of the postwar period the practice has been largely confined to a local level.[4] But larger-scale examples of it have emerged more recently. For example, *Working Lean*, a powerfully moving film by Laura Sky, and *Voices from the Ward*, by a group of York University researchers, are both grounded in personal stories, with experience rather than theoretical hypotheses and positions forming the basis of their authority.[5] In the words of these people's experience, the contradictions between technocratic theory and lived reality come through dramatically. So does the intelligence and commitment of the people participating in the restructuring process, reflecting on what that process is really doing and concluding that it makes no sense from a human point of view.

This story-centred discourse represents an important departure from most unions' involvement in the debate on technological restructuring through the 1980s. Then, most unions largely adjusted themselves to the official corporatist discourse, which was centred on technology as it was currently designed and controlled, to serve the priorities of corporate productivity and market expansionism.

Focused on people, not the technology, this new critical discourse can inform another agenda on restructuring, an agenda formally signalled by unions removing themselves from the existing adjustment agenda and the corporate priorities underlying it. As one example, the Canadian Auto Workers passed a resolution formally rejecting corporate competitiveness as a primary social goal:

"Accepting 'competitiveness' puts us on a treadmill, a rat-race we can't win. . . . It means concessions today and even more concessions tomorrow as other workers feel forced to join the downward spiral. Unlike quality and productivity, the logic of competitiveness adds a dimension that threatens all our achievements."[6]

This discourse can inform another agenda as it reveals and validates the wealth of knowledge and skill that workers bring to the job, as participants in the production/communication process. It attests to the power that working people do have as agents of technology. They're not just objects controlled by it—at least not yet. That knowledge can also suggest new possibilities for people to negotiate more control and involvement for themselves in the lived context of daily work.

Ecofeminist Vandana Shiva coined the phrase "monocultures of the mind" to highlight the colonization of perception that accompanied the imposition of capital-intensive agribusiness in rural India during the so-called "Green Revolution" of the 1960s and 1970s. As a strategy to help resist this monoculture, she calls for an "insurrection of subjugated knowledge."[7] Applied here, this phrase means all the tacit knowledge that people bring to and accumulate through work and living. This shared knowledge is a powerful force that we can use to begin redefining the corporate agenda on globalization and the new economy, and to instead meet our own priorities and our own sense of what's real and relevant.

One of the reasons I admire Innis is that he had a very down-to-earth sense of culture and identity. To him, a healthy culture nurtured the capacity of people to think for themselves—on their own terms—and to act according to their own considered judgement. Self-knowledge synthesized with outside knowledge was essential, in his view. And he championed the oral tradition as a medium for that self-knowledge—and a way of countering dependency on the mass-produced thought associated with the mass media or the official discourse.

The case of the Midland telephone operators offers important lessons here. So far, I've emphasized the importance of a discourse centred in the oral culture of lived experience and the time required to articulate it on your own terms—in other words, to own that experience and utter it (*outer it*, as McLuhan once put it). Although the Midland operators failed in their efforts, there are also lessons to be learned from that failure.

One lesson concerns the structures of communication; the other, the structures of decision-making and governance. It's not enough to have short-distance communications supporting the intimate social relations of local worksites and communities, if the scale on which the official discourse operates is vastly beyond that. For the voices to be heard on a scale large enough to be perceived as publicly relevant, they must be amplified and distributed on a large scale. So networking is a vital part of communication in the new critical discourse, and its organization and funding must be taken seriously. It is vital that people in every community gain the tools not only to tell their own stories, but also to share these in a network. The networking infrastructure and software associated with Solinet and developed by Marc Bélanger at the Canadian Union of Public Employees (CUPE) could act as a base and a model for other networking. Groups such as the National Action Committee on the Status of Women, the Council of Canadians, the Action Canada Network, and the Canadian Centre for Policy Alternatives—some of these are already involved in Solinet—can help share the work and the costs involved.

These structures of communication—designed to be open, inclusive, and locally controlled but also dedicated to sharing knowledge and building a national and even international critical discourse—are as vital as bearing individual witness to the human realities of restructuring in the first place. They are essential to building up enough critical perception to break the monocultural gaze of official reality and to subvert the supremacy of the official discourse, which leaves people anxious about their own and their children's futures but with no sense or confidence that they can do anything about it. These multiple-story and locally grounded lines of perception are in turn vital in the quest to inform a credible public perception of alternatives to the technological restructuring associated with the corporatist agenda. They will help cultivate a broad public consensus and conviction that the power to resist and to rebuild is everywhere, too: in people's voices breaking out of the adjustment mould, breaking the silence on their own experiences of restructuring; speaking out to name their own sense of what's fair or not, what's important or not for themselves, their children, and their society.

A few voices crying outside the frame of the official discourse, outside the window of official reality, are simply cranks, whiners, and misfits. They are voices that don't fit within the frame of what's reasonable and real, and they float away like soap bubbles. But a network knitting these voices together into a coherent social critique and clear statement of what's happening can offer an alternative to cynicism and public despair. Such a network can articulate another agenda for restructuring, and another frame of reference on what's real and relevant: one centred on people and human communities, on justice and equity—not on global competitiveness and cybernetic money markets. And it can support various strategies whereby people come to negotiate the design and governance of the infrastructures and to directly program all the various new tools of communication and production everywhere from the information highway as a whole to the local office, hospital, call-centre, and home workplace.

Ten years ago it might have seemed enough for unions, women's groups, environmental organizations, or groups championing the cause of social justice to lobby governments to implement policies necessary to offset and mitigate the corporate agenda. Since the deregulations of free trade and global restructuring and the general retreat of governments from public policy, it is now up to those of us who reject an exclusionary, cybernetic brave new world to speak up and get involved.

The critical discourse we create must go beyond talk, to direct action. We must define and act on our own agenda for shaping a new global society and economy, in every microcosm where restructuring is happening and in the larger infrastructures and operating systems of the Internet and information highway.

7

On Our Own Terms

If run-throughs are allowed to remain as a managerial prerogative, the men will simply continue to feel that they are victims of technology, inert instruments in a process beyond their control.
–Justice Samuel Freedman

THE KEY ISSUE of our times is not the deficit or social-spending cuts or privatization as such. Nor is it "job creation," broached as if jobs were products to be manufactured as an aside to the globalizing new economy. The issue is the structures of that economy—the infrastructures and operating systems associated with the information highway and all the computerized worksites ramping up and networking around it. Those structures are not opening up opportunities for people to participate meaningfully in the new economy or to expand and keep existing jobs. They're shutting down opportunities for work and participation. This is happening everywhere, in the new sectors of the economy as well as the old, and the results are shifting what had been a fairly democratic distribution of power into a state of dangerous imbalance.

The official theory on technological restructuring as left to business isn't working. The productivity increases of technological change might be increasing economic activity, but this increased activity is not opening up better employment opportunities for most people. Much of the new activity is globally mobile investment, generating jobs in the southern United States, Mexico, South America, and Southeast Asia, though often at starvation wages and under debilitating working conditions. In Canada much of it is taking the form of jobless economic growth: new automated facilities are boosting exports of lumber, paper, minerals, metals, and processed goods, while employment drops or levels off. Furthermore, in the service sector the sophisticated computer software generating many of the new services relies on unpaid consumers (turned into teleconsumers and teleshoppers) to place orders, arrange deliveries, and even complete the final assemblies, but leaves only fill-in-the-blanks scope for personal creativity and involvement.

The deeper issue associated with technological restructuring is this question of *control.* In worksite after worksite where the foundations of the information highway economy are being laid, people are losing control. They are losing the ability to define and program the work to be done, which they traditionally had when the infrastructure was connected to human needs and locally controlled. They are coming to be controlled by the computer operating systems as these systems supply the infrastructure and define the work to be done. In the wraparound, fully programmed work environment in which more employees are now enclosed—in call-centres, in agile work stations, and in telework—people are also being virtually programmed to think in corporate-systems terms. And they're forgetting what "human interface" originally meant—for instance, as public or personal service.

To an important degree, the new computer-defined social relations represent the social contract for the postindustrial era. Such a contract, however, is hardly conducive to an open, inclusive learner and knowledge society. It suggests instead not only a society divided along new lines of power and privilege, but one in which a large segment of the working and consuming population is enmeshed in a set of limited and limiting computer-defined relationships that they can neither negotiate nor significantly control. The mouse and the buttons on the keyboard, like the knobs on a car dashboard, convey the semblance and promise of local and personal control. But on the information highway, in the way it is currently unfolding, most of the real control lies beyond the end-user's reach. It lies in the largely commercial and corporate infrastructures, and operating systems, that control routing, distribution, and pricing (including package discounts on virtually integrated information and service, and the unbundling of comprehensive flat user fees into pay-per-unit billing).

In workplaces, as we've seen, the patterns of control are restricting a lot of people to McJob-like status, working for the computer not with it, needing mental and physical dexterity far more than knowledge to survive. As the patterns of control are extended outward through the virtual economy of the information highway, they will largely determine whether innovative new goods and services offered by Internet entrepreneurs will be widely and remuneratively distributed or not.

The pattern that is emerging in the corporate economy threatens to stifle a knowledge (or leisure) society of meaningful pluralistic participation—the society so enthusiastically envisaged by advocates of the Internet. Freedom is becoming at best push-button multiple choice, except for a carriage-class few. And except for a few, human participation is peripheral to computer operating systems. Service is reduced to functions and functionalities, real and relevant only insofar as they compute. The hard-edged plastic society of homogeneous mass production and consumption is merely being replaced by a softer, more fluid (polymerized) model: more a continuous-feedback loop of computer-animated voice clips and interchangeable packages (of information, jeans, or corn flakes) than a real flowering of diverse humanity. For many of us, then, the promise of real involvement in the postindustrial information society is being amputated and cut back to waiting on the machine. Press enter and stand by.

At heart, then, the struggle to control technological restructuring in worksites is one and the same as the struggle to control the construction and governance of the information highway and the Internet. Both involve a struggle over values: such as equity and reciprocity versus exclusivity and competitive power. Equally, it is a struggle over different social visions: with what Ursula Franklin calls the "growth model" set against the "production model."

In the context of the digitizing new economy and the information highway constructed to support it, the struggle can also be seen as a battle over differing conceptions of communication—as commodity transmission or as community, culture, and personal participation. Within that, it's a struggle between the dominant bias of communication in the modern era, associated with market expansionism, and its opposite, associated with more spiritual values.

The Bias of Communication: Highway Transmission or Communing and Community

The transmission model has dominated communication throughout the modern era, supporting and applying the economy's abiding preoccupation with markets and market expansionism. This model sees communication as a modified form of transportation, the purpose of which is to move information around over large distances as fast as possible. Unlike the community model, it doesn't see communication primarily as a social or cultural process but as product delivery. Communication becomes information transmission, and information becomes a product, a commodity like any other. This is a commercial model of communication, focused on the pipelines and packaging of information, not its content. Its priority is to always expand those pipelines and to keep them full, which also means standardizing the forms of communication for interchangeable (now, multimedia) transmission.

From a policy perspective, supporters of this model interpret content as a separate issue and suggest that policies appropriate for the goals of content can be achieved within an infrastructure devoted to transmission. But the infrastructures are far from hospitable to any communications content: far from it—because of the bias of communication built into those structures. Once in place, that bias is systemic. Harold Innis spent half his career exploring what he called the bias of communication in the modern era. It is a bias towards fast, distance-binding media of communication, media that would facilitate the expansion of markets and of military or commercial empires. This bias, in turn, has influenced how various media of communication are structured and used, which in turn influences what messages they will carry. In other words, as Innis's media studies—from the transportation-communication technologies associated with the fur trade on to the mass-circulation media of chain newspapers—taught him, structures of communication structure consciousness; or, as Marshall McLuhan, Innis's one-time disciple, put it, "the medium is the message." Media structure the social and cultural environments that we live in. While McLuhan focused on the physiological and psychological effects of different media, Innis concentrated on the economic, material structures of communication and their shaping effects. Three of these structural biases or effects are particularly pertinent here in understanding how restructuring for the new economy has also biased developments around the postindustrial "information" society.

One is that cheap, fast, long-distance communication will strengthen the relations between centres and margins and weaken everything in between.[1] This speaks to the organization of the new economy as one dominated by an elite of globally networked virtual corporations. A second is that such communication will centralize decision-making and authority while decentralizing location (work).[2] This speaks both to the new remote-control social relations involved in the new economy and to the widening gap between the haves and have-nots, locally and globally. The third bias says that the faster and further communication travels, the less it can say, and the more its expressive complexity is stripped away—and the capacity of local relations along with it.[3]

The stripping includes not only cultural differences of identity related to specific communities, regions, and groups, but also the minimums of human identity. Applied to modern mass media, the combination of large-scale modern media systems plus commercial sponsorship on which many of these systems depend for their survival and expansion has shaped most programming into an uncritical consumption package. Projecting these biases into the emerging mass media information highway, this is perhaps what Marcel Masse meant when he talked about the new world information technology being "acultural" and "alingual." The ultimate price of universal connectivity might be that we can no longer really speak to each other as people or do much for each other, either. The growing complexity of computer operating codes could be stifling true communication. Perhaps we're being rendered deaf to each other as whole human beings—an ironic reversal of Alexander Graham Bell's initial impulse in telephony, which was to find a way for the deaf to hear. Increasingly, especially when the distancing is augmented by time-metered payment structures or pay-per-use connectivity services and, on the inside, computer-monitoring and performance metrics, the only culture and community left are the digital-bit streams of cyberspace.[4] The structure and pricing biases could mean that the only communication possible, the only language possible, is a digitized assembly of voice clips and cost-conscious functional transactions: as isolated and culturally inert as the sand particles "telling" time in an hourglass.

These biases can only be resisted if the structures (including pricing structures and social relations) are redesigned and renegotiated. And this will only happen if people choose to embrace and uphold the other model of communication with its more spiritual cultural values. U.S. communications scholar James Carey describes the countervailing model of communication as the communitarian or ritual model.[5] Geared more to spiritual than material and commercial values, this model emphasizes communication as process: as social and cultural practice. It emphasizes the communitarian value and meaning of information, the expressive process of sharing information, rather than the forms and containers for transmitting it.

While the transmission model is biased towards fast, light, distance-spanning communications, the community-communications model is geared more towards intimate, face-to-face dialogue and local participation. The transmission model, with its large, capital-intensive scale, favours competition and the centralization of control and power. The communitarian model, which generally depends on less costly, smaller-scale units of technology, encourages reciprocity and a diffusion of control among a plurality of users and an inclusive public space. In the transmission model, time is money. This is a dictum that, through devices such as "call-waiting" and a possible pay-per-minute costing formula for local as well as long-distance communications, can quickly turn even personal phone calls into transactions.

In the community-communitarian model, supported by costless technologies (such as face-to-face conversation), by subsidized technologies (such as the original Internet), or by those regulated to deliver a range of services at an affordable flat rate, time does not equal money. *Time is continuity.* It cannot be metered, or it will lose its integrity as "whatever time it takes" in the context of communication as conversation and communing.

A lot of Canadian cultural practice has emerged out of a mixed-model approach to communications in Canada. The community-communications model has always been subsidized by public money and voluntary commitment. Out of it has come the dialogue-discussion tradition associated with public screenings of National Film Board films—something that, through Studio D, has helped cultivate an informed and engaged women's community in Canada over the past twenty years. Other examples are the old "Radio Farm Forum" and the continuing "Cross-Country Check-up" programs on radio, and the 225 community-access cable channels on television. Finally, consider the vibrant success of public readings in Canada. People don't want more information from the information highway; they want more communication: more dialogue and direct involvement with others.[6]

Public libraries are not only sites for receiving and distributing packages of knowledge and information. They are also knowledge centres in their own right. A library's role is to gather knowledge that is both locally distinct and relevant. Libraries also help people inform themselves on their own terms, which aren't necessarily the same as the standardized and often Americanized terms provided by a few Wal-Mart-scale data bases.[7] If Canadian public libraries "ramp up" to the infrastructures of the information highway on their own terms as communitarian and cultural institutions, the highway infrastructure will have to be negotiated accordingly. This means locating a lot of effective communicating and programming capacity in the hands of end-users, extending librarians' jobs and the freenet model of inclusive community information services, and preserving the inclusive, open-architecture infrastructure associated with the Internet.

The infrastructures and operating systems of the information highway must be organized so the highway becomes an extension of existing communities, of the diversity of communities—grounded in particular geographic, ethnic, and sometimes specialized knowledge—and not the other way around. Otherwise, these community institutions could be hollowed out as social and cultural centres, becoming little more than on-ramps and off-ramps to remote information highway information-file service centres. Except for a privileged, niche-market few, libraries could become mere gateways to vast information-processing and transmission systems. In education, learning could be reduced to fill-in-the-blanks, multiple-choice computer-interactivity developed and delivered by information-service industries. Similarly, hospitals could become gateways for more and more contracted-out remote services, from sophisticated diagnostics to inventory and personnel management.

Anything and everything digitizable will move, keeping the transmission lines full. Self-employed Internet entrepreneurs will still be free to write creative, customized software and to offer innovative information goods and services. But like Canadian filmmakers, who are structurally frozen out of the country's main distribution channels, they might have only the residual transmission capacity on the information highway for trying to communicate with an audience and make a living from their work—whatever is left over by the big virtually integrated information systems and service suppliers such as Microsoft, Time Warner, General Electric, Rogers, and members of the Stentor consortium.

Significantly, both the Canadian Library Association and the Ontario Library Association have stressed participation rather than mere technical access as a key policy principle for the information highway. Participation suggests the community-communications model, while access connotes the transmission model. Unfortunately, the 1995 report of the Information Highway Advisory Council (IHAC) doesn't call for a balance between the transmission model and the community-communitarian model of communication. At the level of rhetoric—including its title, *Connection, Community, Content*—the report hits all the right, feel-good tones. But in its substance it clearly advocates a commercial transmission-model approach with a limited government regulatory role.

The Council defines fairness and sustainability only in terms of the commercial transmission model. Its approach ignores the possibility that confining communications to a competitive transmission model of development might not in itself be fair or sustainable. It also reduces the larger ethical values of fairness and sustainability to a subset of narrowly defined technical and economic values. It discusses sustainability as "sustainable competition" and fairness as fair access to the technology.[8]

The IHAC report articulates only one side of the debate. It presents only one negotiating position: a corporate agenda, with its ideological basis disguised as economics and neutral technics. The other agenda, with its fundamentally different values and vision, has to be retrieved. A policy framework for accommodating and applying both visions has to be negotiated and funded.

By the mid-1990s the bulk of public spending on the information highway had gone into infrastructure and value-added technologies such as network-management software. In other words, public money is largely going towards building the transmission model of communications. The community-communitarian model requires equal funding. The corporate commercial economy must remain integrated with the social and community economy, and a balance of power has to be maintained between them.

A Social Contract for the Information Highway

Participation is perhaps the first priority of a social charter or contract for the information highway and for building a truly new economy that would include both market and social/community priorities.

The Ontario Library Association has defined participation as requiring "a guarantee of the right of every individual to ready, effective, equitable and timely access to information in order to participate fully in the social, economic, political, educational and cultural life of the country."[9] In the same vein, the Public Interest Advocacy Centre has defined "basic and essential service . . . as that level of service required for full participation in society."[10] A report summarizing consumer and cultural groups' submissions to the Information Highway Advisory Council during its brief public hearings defines interactivity not in the transmission-model terms favoured by corporate communications carriers—that is, as technical functionality—but in more community-cultural terms—as communicative interactivity.

What this means, according to communications consultant Andrew Reddick, is "a user-controlled, fully switched two-way system of communication which allows users to be information providers and participants as well as consumers."[11] This approach doesn't view the emerging global communications networks as pipelines for delivering commodities and distributing workloads, but as a new context for living and working to the fullest extent of one's talents, skills, and interests. If this environment is to be democratic and richly human, all the users within it need some scope for programming and controlling the technology themselves so they can program reciprocity and inclusive participation into its systems and infrastructure. Currently the structures' controlling/creative power is being concentrated among a few large media and information-service providers, which means that the highway is becoming the equivalent of "six lanes coming into the house, but only a bicycle path leading back out."[12] In other words, in the way the networks are being structured and governed now, relatively few people will get to be creative participants on their own terms; most will be teleworkers or teleconsumers. Relatively few people will get a chance to use the information highway as an extension of their minds and imaginations; most could become the extensions of a few corporate minds associated with computerized production, learning, or other services. And only a fraction of the rapidly growing ranks of the self-employed will be able to make a living by finding a hundred clients for this computer program, two hundred users for that information package, and so on. A few might make a killing, while most will have to settle for McJobs.

The critical discourses on technological restructuring and the information highway are really one and the same, and participation is the principle uniting them. That principle is crucial for the future scripting of the information highway and for the renegotiation of the technological restructuring that is leading us down that road. The new social relations currently being laid down in our computerizing institutions—the "new cybernetics of labour"—will be decisive in determining whether or not the information highway is a site for pluralistic participation, a place where people can fully employ their knowledge, skills, talents, and ambitions. As things now stand, those relations have to be urgently renegotiated to ensure participation.

The core issue is the distribution of power in society, and the need to restore and preserve a balanced distribution in the technological infrastructure and operating systems of the new economy. If justice is to be served through inclusive participation in the information highway and/or full and meaningful employment in the new economy, the power to operate and program these structures must be broadly and democratically distributed. Power must be balanced between those who build and manage the systems and infrastructures and those who use them.

This is what Justice Samuel Freedman of Manitoba meant when he recommended that technological restructuring be negotiated and that the tradition of managerial control over technology be abolished. Freedman realized that technological change in the late industrial period went far beyond the issue of changing tools of production. It was altering the larger social environment, including the basis for making a living. He argued that in a society that called itself just and democratic, it was morally wrong to treat people as simply a labour "commodity," to be arbitrarily dumped as redundant. "What is required if the men are not to feel that they are victims of a plan instead of participants in it is negotiation on the basis of parity," he concluded.[13]

In the past decade the ability of unions to protect jobs in the face of technological change has become more diminished than ever. In March 1995 the Liberal government legislated thirty thousand rail workers back to work after a strike-lockout over, among other things, the negotiated right to job security in the face of technological restructuring.

Liberal Labour Minister Lucienne Robillard defended the back-to-work legislation in the Commons, saying, "The best job protection employees can have is the certainty that their employer can face competition."[14]

But she is wrong. In an inclusive democratic society it is morally wrong for people to be reduced to digitized dice in a corporate monopoly game. What the Manitoba judge said in his 1965 inquiry into technological restructuring in one of Canada's oldest transportation-communication systems vitally applies to the structuring of Canada's digital transportation-communication systems today. The right to negotiate the new networked context of their work and working would surely have helped the Midland telephone operators, for instance. It would have given them the power to negotiate the restructuring of the local telephone exchange into both an efficient call-processing system and an effective medium for community communication and local information services.

Similarly, it would also have helped a group of Vancouver municipal office workers who in the late 1980s came up with a way to computerize city hall that would have both automated much of the routine administration (as planned) and extended their jobs even while extending the scope of service to the public. In a participatory research project spread over several months and including technical experts as resource support, the Vancouver Municipal and Regional Employees Union set out to inform the office workers about the computerization plans being developed by management and its consultants. But the union also worked to involve the employees as "experts" in developing an alternative plan, drawing on their own tacit knowledge and work experience. The plan differed dramatically from management's plan in that it was not centred on the systems logic of running city hall like an automated system—the transmission model of communication. Instead it was centred in the logic of the context—in the community services that city hall workers provided to the public through various systems.

In one proposal, the project suggested that switchboard operators who had simply connected calls to various departments could become information and referral agents, linked to a data base detailing all public services—even those outside the municipal sphere—an idea which was successfully implemented by the United Services Auto Association, not automating work but "informating" it.[15] Another suggestion was for the development and continual updating of a registry of "at risk citizens"—people just released from hospital after surgery, newborns and their mothers—which would be of great use to a number of local social-service agencies. Yet another called for creating and maintaining an electronic inventory of trees on all public lands throughout the city. The public works people could consult this before laying down new roads or water pipes.

The plan gave the municipal workers a negotiating position on technological change beyond simply adjusting to or resisting management's agenda on restructuring. It gave them a position that was neither for nor against technology as given, but for their own design of technology and their own sense of the goals it should serve in the particular context of Vancouver as a living community. They hoped to negotiate a compromise on technological restructuring with city hall management. But management refused to negotiate, and the project went no further than a final report authored by the bold and creative people involved.[16]

The report noted that overcoming the mystique of the expert and the corresponding low self-confidence among the worker-participants was the biggest obstacle of the whole project, and was never fully overcome. "Many people found it difficult to value their own ideas or potential for contribution," the report said. "They believed their suggestions would not be seriously considered by management." The employees "would likely participate more readily in management-initiated meetings," because that way, they believed, "their ideas have a greater chance of coming to fruition."

The official discourse on technological restructuring, centred around the needs of the corporate economy and the voices of its technocratic managers, can be immobilizing, as this case makes clear. Still, it is possible to articulate alternatives, as this case also makes clear, and taking the time to do this, in worksites everywhere, is vital.

The situation is still fluid. Canadian libraries and hospitals, as well as Canadian education, which is undergoing a similar transformation, can be renewed around an ethic of public service, if the technological restructuring of these institutions and the computerization of the services involved are defined by the people who are directly involved in giving and receiving those services, and who are dedicated to health or education as a public good, not as a brand of commercial business. The Canadian tradition of seeing health, education, and culture as public resources can be renewed if these people can control the choice and use of technologies in those spheres and negotiate the restructuring process in a discourse of direct experience and action.

This discourse doesn't need grand plans to get started. Perhaps it's enough if people can see how things could be done differently, serving priorities other than faster, cheaper equals better, and simply start doing it. In workplaces with quality circles, total quality management, and similar "worker-empowerment" programs, employees could take some of the time they spend meeting, often on their own lunch breaks, to define the meaning of quality in their own words. This could include the quality of their lives, both as part of a working community inside the hospital, office, or factory and in the larger community of family, friends, and neighbours. It could also mean quality in terms of the service or product they are employed to produce.

The CAMI automotive factory outside Ingersoll boasts a sophisticated computer-integrated manufacturing system for making cars customized to meet customers' individual needs. Yet the cars and jeeps coming off the two production lines look almost completely identical: the same profile, the same basic structure and design. When I toured the plant I looked down that line of undifferentiated vehicles and thought, what a waste.

People come in very different shapes and sizes, with very different needs in automobility. If the highly computerized work environment of today's car factories could be run on truly participatory principles, with workers able to participate fully in customized production on their own terms, and able to direct the use of new technologies to serve those terms, perhaps they would use these powerful technologies to truly customize cars to meet the widely different needs of people. For example, why can't the seat-carriage mechanism be customized for people with physical disabilities—so the seat could slide out from the passenger or driver's side to make it easier for someone transferring from a wheelchair? Why can't small vehicles be outfitted with seats that can change into beds the way they do in family-size vans?

I'd like to think that if the Vancouver city hall workers were conducting their project today, they would contribute their ideas, knowledge, talent, and even the technology they work with to support and cross-subsidize a local freenet. I'd also like to believe that if they just went ahead and did it, they could enact a mixed-model approach to communication as both technical functionality *and* public service. There is an alternative to information services being reduced to computer-mediated voice clips. The information highway can be an extension of existing human communities, not a silicon strip mall and cybernetic workplace.

To build and apply this vision requires a concerted networking effort. It requires electronic and other networks to construct a critical mass of public perception that reaches out and challenges how the new economy is being structured—that says we can restructure it with participation and the needs of the social environment at the centre, not downsizing and deficit-cutting for a globalizing corporate economy operating on the lowest common denominator. It also requires a new de-institutionalized form of organization in unions and other social-change organizations.

Virtual Unions and Coalition-Building

Unions need to remember their roots as community benevolent and self-help societies. In the de-institutionalized 1990s, these roots would suggest a reversal of the hierarchies that developed in the heyday of bureaucratic business unionism. The new critical discourse on restructuring needs to be centred in actual experience and concrete actions that will turn the restructuring agenda around in the specific sites of the here and now as well as in the larger policies of the information highway. It requires therefore that people largely speak for themselves, instead of being spoken for by union leaders and professional staff. It also implies that people act on their own behalf, with staff working to support their discourse of action.

Solinet could serve as an E-mail and electronic conferencing medium, uniting local groups of working people as they critique their specific situation and figure out what they can do about it. It could provide not only a local communications medium, but also virtual links to various experts and knowledge bases, as well as to other institutions—everything from the Canadian Labour Congress to the National Action Committee on the Status of Women to Greenpeace—to create a common front or "virtual coalitions." In the postindustrial, postmodern era, process and relationships predominate over fixed structures. Contemporary institutional structures are digital, and virtual institutions can be created at the push of the "send" or "enter" button, allowing people to share files, work jointly, and nurture solidarity through daily contact across town, across the country, and even around the world.

Any participation in the "official" discourse on restructuring and the information highway should be organized to function as an extension of this grounded critical discourse. This suggests too that anyone representing workers and communities should spend at least half their time within a concrete work situation or community, so they can always stay connected to its essence and relevancies and work to effect change in the here and now.

In a second shift of focus, the staff of unions and social-change groups could perhaps also spend more time extending the lines of virtual communication among de-institutionalized and de-unionized working people, so those people can speak to each other and create virtual coalitions around common concerns—such as working hours, working conditions, benefits, and tax policies.

There should be an E-mail account and electronic self-help available to all teleworkers. This could be offered by Solinet directly, or indirectly through community freenets, social-interest groups such as the Public Interest Advocacy Centre, or women's groups such as NAC and CCLOW (The Canadian Congress for Learning Opportunities for Women). Similarly, unions and social-change groups can offer virtual services such as training and job counselling and enlist individuals and work groups as virtual members of project-specific coalitions.

Coalition on working hours: In 1994 the number of Canadians—mostly men—working fifty hours a week and more jumped to 22 per cent, from 17 per cent in 1976. Meanwhile, a growing number of other Canadians—and especially women—are only able to get part-time jobs, if they can get jobs at all. To make ends meet, more and more people are moonlighting, taking on two or sometimes three jobs.[17]

This is the reverse of the promised leisure society. In fact, while output per hour of work has doubled since 1945, leisure time has dwindled.[18] Not only is this threatening people's health: look at the word *over-time.* More and more people are having to work over what is a normal allotment of time for work, over what is conducive to a balanced life of work, family, and leisure. And at the same time many other people aren't getting a chance to work enough to make ends meet, or even to work at all. It's been estimated that if Canadian work hours could be reduced to the average levels in Germany, up to two hundred thousand more people could potentially be employed full-time.[19]

There is tremendous resistance to changing this situation. Some is built into systems such as income tax, unemployment insurance, and other benefits. A Statistics Canada study attributes much of the polarization in working hours and wages to both the large difference in benefit costs between hiring full-time and part-time people and the shift away from progressive income taxes since 1984, which made it more attractive for people with full-time jobs to work longer hours, thus allowing employers to avoid hiring another person full-time.[20] As well, there are rigidities in the cost of living once people buy a house or get used to owning a second car, home-entertainment centres, or being members of sports and fitness clubs.

Some resistance can also be traced to the masculine mystique of our society. This includes the equation of money with power and more money with greater power and prestige. But it goes beyond this to include sexual politics around who picks up the kids, makes supper, and does bath duty versus who has to work late. There's a certain machismo factor built in there too, which has been documented historically in W.K. Kellogg's experiment with a six-hour day at Battle Creek, Michigan. Kellogg's Corn Flakes factory turned three eight-hour shifts into four six-hour shifts, added 25 per cent more workers to the payroll, and increased hourly wages by 12 per cent. Kellogg took this step in 1930 as a way to ease the local unemployment situation. Some fifty years later the six-hour day was popular only among the women, who saw it as giving them more time for family, community, and genuine leisure. The men repudiated the experiment, seeing leisure as something "for silly women or sissy men," according to U.S. historian

Benjamin Hunnicutt. In the end senior male workers combined with Kellogg management to lobby against the shorter work day, and it was dropped as an option in 1984.[21]

It might help if we could resurrect an idea from the late nineteenth century called the Nine-Hours League.[22] The league, which anyone could join whether they were a member of a union or not, was one of the first examples of coalition-building for decent labour standards, and it worked. If such a coalition were formed today, it might break down the solitudes between the unemployed, the underemployed, and the overemployed. If it also included community, environmental, and women's groups, such a virtual coalition could help dramatize the context in which this issue is played out and the importance of stable employment and a standard workweek. It would also dramatize the role of individual choice and responsibility in shaping and reshaping the restructuring agenda—and the importance of time within this phenomenon: how we use our time, whether we're driven, or whether we actively resist being driven and choose a leisure (and conserver) society instead. Our personal choices count.

Labour standards: Another coalition is needed to develop common minimum labour standards for part-time, term, and other contingent workers so they are not treated like post-it-notes, stuck on one moment and thrown away the next, with no chance for continuity or in-depth involvement. At a minimum, all part-time workers should qualify for unemployment insurance. They should also be entitled to training programs and the chance to grow on the job.

A labour-standards coalition could help draft a minimum-work contract for the growing numbers of people taking term or contract work in the form of telework out of their homes or satellite offices, and work to have it respected and enforced. The Writers' Union of Canada and ACTRA might serve as examples here, both as prototypes of virtual unions for de-institutionalized and self-employed people, and for developing labour-standards policies on behalf of their members. For example, the writers' union developed a standard writer-publisher contract in the early 1980s and spent the next ten years negotiating to have its clauses on royalty rates and mutual obligations accepted by publishers operating in Canada, which by and large has happened.

Machine census and "head tax": Machines and the scale-up of machine technology have been at the heart of the centralization of power throughout the industrial era. In the latest scale-up, associated with computerization, the information highway, and the new economy, the machines have become more powerful and people power—certainly in the corporate "formal" economy—is on the wane.

In trying to restore the kind of balance of power that is required in a democratic and inclusive society, it is essential to dramatize the presence of the machines. The downsizing of people in corporations is often paralleled in the increasing use of computers, robots, and networked, automated systems. People are being displaced from their own economy, and machines and machine intelligence are taking their place. This trend needs to be seen in order to be publicly debated.

One way of highlighting this situation would be through organizing a citizens' census on computers.[23] Such a project would quantify the explosion of computer-related technologies that has been occurring since the 1970s, with no let-up in sight. It could also quantify disparities in budgets—for instance, the federal government cutting forty-five thousand jobs while continuing to spend close to $4 billion a year on information technology and related services. What is the equivalent budget for high-tech medicine and related information systems in hospitals, which are now being hit by spending cuts and related "downsizing"? The census project could provide people with an initial reason to network from one worksite to another as they used Solinet and other communications media to gather, collate, and analyse the information. It could also generate an initial basis for rethinking technological restructuring by asking whose priorities are really being served by it: the people selling high-tech systems or the public needing health care and those with skills and a commitment to provide it? Asking ourselves *whose benefits* rather than *what benefits* would provide a perspective from which we could ask a number of questions. Is the escalating spiral of computerization scripted simply to keep the high-capacity communications lines of the information highway full? Is its main goal to keep the new computer-communications and information-media-systems conglomerates in business and to deliver economic renewal to the corporate economy at the expense of people and human communities? Is the exploding memory capacity of microchips some monstrous rebuilding of the Tower of Babel, full of bug-plagued digital babble where once there were people dealing with each other in communities of health care, learning, and other public services?

A kind of "head tax" on computers would help to dramatize the politics behind the choices involved. While the infamous and racist "head tax" imposed on Chinese immigrants in the early part of this century restricted the entry of Chinese into Canadian society, a head tax on computers would check the rampant proliferation of these machines, which current write-off and deduction provisions in tax regulation encourage. The Tobin tax is a variation on this idea. Tobin proposes a tax on foreign-exchange transactions, which occur at a rate of $100 billion a day in New York alone.[24] But outright licence fees and escalating taxes on digital information flows would more broadly address the power politics involved. Such a step would also be consistent with some of the earliest measures adopted to police machine technology. In the late eighteenth century the government in England introduced a number of laws restricting the proliferation of mechanical looms. The legislation didn't forbid the new technologies. It merely regulated them so that the scale associated with the new mechanical looms and the commercial entrepreneurs rushing to invest in them wouldn't crush the traditional household-based industry, deskilling the work into jobs for children at below minimum wages while destroying local economies and throwing thousands of weavers out of work. For example, at one time clothiers were prohibited from owning more than one loom. They were also forbidden to put out looms for hire—that is, to separate the (capitalist) management of looms and weaving from the work itself.

However, as the power of the new mercantilists and related manufacturers grew, so did their lobbying power. Over a few years in the early nineteenth century, the roughly seventy laws that had regulated the new technology were overturned, and that's when the Luddite movement was born. It emerged because of this deregulation, this switch from public regulation of technology in the interests of the common good, to strictly private regulation and control by business. Then, as now, the issue was not to destroy the technology, but to bring its use back under public control so it would serve the needs of the larger economy of life, not just the needs of the new commercial industrial economy. Based on the historical evidence, the Luddites "saw a role for capital in society but within limits." They "held an essentially moral view of economic relationships, a view . . . which harked back to the old paternalistic concept of stability and regulation," which, to them, was at odds with "the amoral economy of the innovators."[25]

The issue is as urgent now as it was two hundred years ago: regulating limits on the proliferation of information technology, so that instead of eliminating people from the economy and cheapening what they do, the technology includes people and extends what they can do.

The "reverse strike" and workfare: The idea of a "reverse strike" rests on the moral right of people to work. It also rests on their right to define what is meaningful and relevant work. As an historical event, staged by unemployed workers in the Sicilian town of Partinico in the mid-1950s, it also required the moral courage of people to act on their self-described work agenda. In Partinico, in a miserably wet December of 1956, that work involved repairing the road into town. Apparently, the road had deteriorated to the point that, in the autumn rains, it had become nearly impassable. The people wanted it repaired, but the government wouldn't hire people to do it. Under the leadership of pacifist social activist Danilo Dolci, the people seized the initiative and started repairing the road themselves. Within an hour of their starting to work, however, the police had forcibly stopped them and arrested Dolci.

On a certain level, therefore, the action was a failure. Yet many people remember it as an example of citizen resistance to the arbitrary governance of human community, contrary to human needs.

Dolci defended himself at his trial on charges of obstructing authority: "If you were to order me, Your Honour, to kill my father . . . I would not obey you, because my conscience would not let me; because it is a crime. For me, it is just as much a crime against the land and against mankind to stop work—for the very good reason that it is our duty to work, not only for ourselves but for the good of all. And besides, too many promises had been made, year after year, and then never kept."[26]

The reverse strike idea is a dangerous one, in that seizing the moral initiative to work without having the means to pay people decently for the work they do can invite them into a martyrdom that simply reinforces the status quo. Despite the risks the idea might be worth exploring—by, for instance, the forty-five thousand federal public servants who have arbitrarily been declared "surplus" to the job of providing Canadians with social and other public services; by the ten thousand employees being laid off by Bell Canada; by the hundreds being "downsized" by the Canadian National and Canadian Pacific railways; or the hundreds more being cut from health care, banks, and other financial services. The reverse strike might also be of interest to the community-service organizations that are being conscripted as partners in the Ontario workfare program without even having been consulted.

Acting on the reverse-strike idea that people have the moral right to define what work is necessary and to do it, these community organizations could work in solidarity with Ontario welfare recipients and create a community dialogue around the work that should be done. Instead of being a meaningless make-work program or a species of forced labour, workfare could be redesigned as meaningful work and working experience, negotiated by the participants. For instance, welfare recipients might decide to work on developing local community freenets, through the installation of user terminals in laundromats, and extending their uses to include, for instance, electronic scheduling and bulletin boards for volunteers with the local Meals on Wheels project. Some valuable new communications services might emerge from these efforts, and even some new virtual businesses, possibly funded as joint-stock co-operatives involving various not-for-profit groups that formerly depended on government grants.

A Constitution for the Information Highway and the New Economy

There is also work to be done at the macro level of policy-making and in the discourse there. We need a social charter or constitution for the information highway. We need a social and cultural agenda that defines fairness and sustainability in terms of people and a diverse, inclusive social environment—not solely in technical terms such as access to the highway infrastructure. Such an agenda could be developed through a consensus conference or constituent assembly involving all the social and cultural groups that submitted briefs to IHAC during its policy consultation process, which would reconvene over a number of years to fully articulate this agenda and to see it implemented. Its purpose would be to articulate a communitarian model of communications that would address the twin goal of full employment and full participation in the new economy. But to succeed, such a conference would have to be organized and controlled by these groups as well, not by corporatist government bureaucrats who would otherwise control the agenda, the frame and terms of reference, and, finally, the wording of the resultant consensus.

Existing coalitions such as the Alliance for a Connected Canada and People for Affordable Telephone Service could take the initiative here.[27] But they must trust the communication-as-community-building process and take the time necessary to cultivate an inclusive community-based discussion. This might involve an initial stage of groups cultivating dialogue at their annual meetings, through newsletters, or on members' E-mail networks, and returning the discourse to these networks after the conference or assembly.

Having named a people's agenda for full and democratic participation in the information highway and the new economy, the people and groups involved in this discourse must also act on their work. This could include lobbying individual members of parliament for their endorsement, much like women did in 1981 to have gender-equality rights included in the Canadian Charter of Rights. That year, in February, an ad hoc committee of women and women's groups had staged a consensus conference on women and the constitution, which led to the equality clauses (notably Section 28). And if it hadn't been for the dedicated follow-up lobby, they could have easily been ignored by the government.[28]

The various groups and institutions rallying behind this participatory design plan for the new economy and its infrastructure must also insist on negotiating policies based on its values and vision. This means, where possible, working out a mixed-model approach to communications policy around the information highway. In other words, ensuring that society exists both through communication and in communication, with people sending and receiving information and talking with each other.[29] This means accommodating corporate systems goals of technological expansionism with community and public-service goals where possible. It also means, where necessary, curbing and regulating business so that its systems serve social priorities *now*, not through some trickle-down, after-the-fact effect.

The information highway infrastructure, having become the axis of the new economy and, to a large extent, of social and cultural life too, is too important to be left to big business. It should probably become a public utility; although at arm's length from the government and as much as possible governed through networks of local boards that could interpret national and even international standards and principles in the diversity of local contexts.

A coalition or common front for full employment and full participation in the new information economy would also address the deficit problem by restoring the tax base, which has been lost to unemployment and underemployment in the past ten years. That way, we can renew the social programs that are essential to a just and healthy society.

Turning restructuring around depends first and finally on the committed participation of people themselves, acting at times individually and, perhaps more effectively, in groups and in virtual unions and coalitions, making a difference in the daily context of technological practice—in this agile factory, in that otherwise isolated call-centre, in this hospital and in that supermarket, as well as in government at all levels. In this era of de-institutionalized governance, it is up to us—all of us—to take the initiative and to reconstitute ourselves as an inclusive, caring, just, civil society.

I've never entirely trusted the futurist slogan, "think globally, act locally." I think you can network globally, but to really know what you're doing, to have the strength of your own convictions—that what you're doing is appropriate locally and personally—you have to think locally and act locally too. The new critical discourse on technological restructuring and the information highway requires local, regional, and even global networking. But it must be grounded in local dialogue, local community-building, and locally appropriate action. It must run from telling it like it is in specific working and living situations to looking at what we are trying to do about it and at what kind of postindustrial society we want for ourselves, our children, and our friends.

Regional, national, and international networking is vital as a source of sharing analysis and gaining strategic support through common and collective action. But its strength rests entirely on the vitality of the local talk and action. Everything I've read and personally experienced through twenty-five years of "grassroots" activism in the women's movement and the peace movement convinces me this is true. We are the movement. We are the information highway and the new economy. We have to act as though these structures are a place where people really are the bottom line, where communication is first and foremost a matter of human relationships and community. By acting that way, we can make it so.

Notes

1. Behind the Silicon Curtain: Perception Management and the Adjustment Agenda

1. Ekos Research Associates, *Rethinking Government*, Ottawa, 1995, p.90.
2. *The Outsiders: A Report on the Prospects for Young Families in Metro Toronto*, Family Service Association of Metropolitan Toronto, 1994, p.4.
3. Goss Gilroy Inc., Management Consultants, "Impacts of the Information Highway on Employment and the Workplace," Ottawa, 1995, p.61. Rather than "breadwinner," Statistics Canada uses the term "primary earner," meaning whoever in the household is earning the most money.
4. Deloitte and Touche study, quoted by Sandra Rubin, "Banks Discount Talk of Massive Job Losses," *The Ottawa Citizen*, Sept. 22, 1995.
5. These figures are from the federal census, under the heading "Protective Service Occupations—Guards and Related Security Occupations."
6. Margot Gibb-Clark, "Jobs Keeping Couples Apart," *The Globe and Mail*, June 2, 1995, p.A1.
7. "Workplace Called Hazardous to Health in Stats Can Report," *The Globe and Mail*, Oct. 31, 1994; David McKie, "A Problem Journalists Can Ill Afford to Ignore," box insert in Vivian Smith, "Repetitive Strain Injury Can Get Really Awful Really Fast" (cover story), *Media* (The Canadian Association of Journalists), Vol.1, No.2. (July 1994), p.11.
8. Pat Armstrong, Jacqueline Choiniere, Gina Feldberg, and Jerry White, "Voices from the Ward," *Take Care: Warning Signals for Canada's Health System* (Toronto: Garamond Press, 1994), p.65.
9. Nicholas Chamie, quoted by Bruce Little, "Competition, Technology Cited as Causes of 'Jobless Recovery,'" *The Globe and Mail*, Oct. 5, 1995.
10. "Federal Expenditures on Information Technology," document prepared by Treasury Board for Office of the Auditor General. Figures obtained by the author through the Library of Parliament, Ottawa.
11. John Partridge, "National Trustco Axing More Staff," *The Globe and Mail*, Oct. 11, 1995.

12. Theresa Johnson, "Information Superhighway and the Consequences for Work," *TECHnotes* (Ontario Federation of Labour), Vol.10 (June 1995), p.10. The original statement was: "The tendency is to replace the human infrastructure . . . and to invest in technological infrastructures."
13. Robert Brehl, "Get Set for $4 Rate Hike," *The Toronto Star*, Nov. 1, 1995.
14. "Third of People Yearn for Information Highway, Poll Finds," *The Globe and Mail*, Oct. 15, 1994.
15. Arthur Cordell, "New Taxes for a New Economy," paper presented at "Scenarios for the Future: Work and Education" conference, University of Toronto, Sept. 14, 1995, p.4.
16. The term "cyberspace" was coined by William Gibson in *Neuromancer* (New York: Avon Books, 1984), p.5.
17. Ursula M. Franklin, *The Real World of Technology* (Toronto: Anansi Press, 1990), p.24. Franklin refers to "a culture of compliance."
18. John Ralston Saul defines corporatism as a form of government based on involved experts rather than citizens. Being citizens in a society at large becomes secondary to being part of professional and expert groups—corporate groups with which the state bureaucracy consults in formulating policy. Saul defines this quasi-fascist form of government as the real threat to democracy in our times, and I agree. See John Ralston Saul, *The Doubter's Companion: A Dictionary of Aggressive Common Sense* (Toronto: The Penguin Group, 1994), pp.74-79.
19. Marcel Masse, presentation at "A World beyond Borders" conference, Canadian Conference of the Arts, Ottawa, 1994, p.17.
20. See "Time, Moore's Law, and the Technological Dynamo" in Chapter 2.
21. "The Proven Path to Outperformance," large glossy ad for Trimark Mutual Funds included as insert, *The Ottawa Citizen* and *The Globe and Mail*, Jan. 24, 1995.
22. Marshall McLuhan, "Living at the Speed of Light," *Maclean's*, Jan. 7, 1980, p.32.
23. Stephen Dale, "Who's Quoted Most?" *CCPA Monitor*, June 1995, p.8.
24. Dorothy Smith, *The Conceptual Practices of Power: A Feminist Sociology of Knowledge* (Toronto: University of Toronto Press, 1990). I am drawing from the wisdom of the whole book here, but in particular, passages on pp.4, 17.
25. John Ralston Saul, *Voltaire's Bastards: The Dictatorship of Reason in the West* (Toronto: Penguin Books, 1992), p.30.
26. James Carey, "Canadian Communications Theory: Extensions and Interpretations of Harold Innis," in *Studies in Canadian Communications*, ed. Gertrude J. Robinson and Donald F. Theall (Montreal: McGill Programme in Communications, 1975), p.45.
27. Rick Salutin, "The Picture Outside the Frame," *The Globe and Mail*, March 10, 1995.

28. Marshall McLuhan, *Understanding Media: The Extensions of Man* (New York: McGraw Hill, 1964), p.202. "The advertising industry is a crude attempt to extend the principles of automation to every aspect of society. . . . It stretches out toward the ultimate electronic goal of a collective consciousness."
29. Interview, December 1994.
30. Jacques Ellul, *Propaganda: The Formation of Men's Attitudes* (New York: Vintage Books, 1973), p.74.
31. These are some of the prizes and rewards offered for good team performance at the CAMI automotive assembly plant in Ingersoll, Ontario, a joint venture between GM and Suzuki. Plant operations are based on Japanese management programs such as "kaisen," "taien," and "total quality management."

2. *The Chip and Programmed People: The Real World of Restructuring*

1. Thanks to Nicole Smith, a former student of mine in Ottawa, for making this point.
2. CBC Radio News, 8 a.m., Sept. 28, 1995.
3. Samuel Freedman, "Report of the Industrial Inquiry Commission on Canadian National Railways 'Run-Throughs,'" National Library, Ottawa, 1965, p.95.
4. The Honourable Michael Wilson, Minister of Finance, *A New Direction for Canada: An Agenda for Economic Renewal* (Ottawa: Department of Finance Canada, 1984), p.49.
5. Linda McQuaig, *Shooting the Hippo: Death by Deficit and Other Canadian Myths* (Toronto: Penguin, 1995).
6. Diane Bellemare, *What Is the Real Cost of Unemployment in Canada?* (Ottawa: Canadian Centre for Policy Alternatives, 1994), p.12.
7. Canada, House of Commons, Standing Committee on Human Resources Development, *Minutes*, Issue No.5 (March 7, 1994), p.22.
8. Saul, *Voltaire's Bastards*, p.377.
9. Langdon Winner, "Do Artifacts Have Politics?" in *The Social Shaping of Technology*, ed. Donald MacKenzie and Judy Wajcman (Milton Keynes, England: Open University Press, 1985), p.28.
10. Elaine Bernard, "Science, Technology and Progress: Lessons from the History of the Typewriter," *Canadian Woman Studies/les cahiers de la femme*, Vol.5, No.4 (1984), p.13.
11. Michele Martin, *"Hello Central?" Gender, Technology and Culture in the Formation of Telephone Systems* (Montreal and Kingston: McGill-Queen's University Press, 1991), pp.21-22.

12. Ibid., p.98.
13. James Gleick, "Making Microsoft Safe for Capitalism," *The New York Times Magazine*, Nov. 5, 1995, p.55.
14. Ursula Franklin, *The Real World of Technology*, p.31.
15. Sam Sturnberg, "Jobs, Jobs, Jobs," *TECHnotes* (Ontario Federation of Labour), Vol.10 (June 1995), p.6.
16. See Donna Haraway, *Simians, Cyborgs and Women: The Reinvention of Nature* (London: Free Association Books, 1991), p.161. Haraway appears to have coined the phrase. The elaboration of what it means is my own.
17. Gordon Betcherman, Kathryn McMullen, Norm Leckie, and Christina Caron, *The Canadian Workplace in Transition: The Final Report of the Human Resource Management Project* (Kingston: Industrial Relations Centre Press, Queen's University, 1994), p.82.
18. "Labour Market Analysis Data," document, Analytical Studies Branch, Statistics Canada, Ottawa, 1994, p.29.
19. John Myles, Garnett Picot, and Ted Wannell, *Wages and Jobs in the 1980s: Changing Youth Wages and the Declining Middle* (Ottawa: Statistics Canada, Social and Economic Studies Division, 1988).
20 "Labour Market Analysis Data."
21. R. Morissette, J. Myles, and G. Picot, *What is Happening to Earnings Inequality in Canada?* Business and Labour Market Analysis Group, Analytical Studies Branch, Statistics Canada, Ottawa, 1993, p.2.
22. D.W. Livingstone, D. Hart, and L.E. Davie, *Public Attitudes Towards Education in Ontario* (Toronto: OISE Press, 1995), p.27. Personal conversation with David Livingstone, Feb. 1996.
23. Canadian Press, "GM Job-seekers Need High School," *The Vancouver Sun*, Jan. 13, 1995.
24. Graham Lowe and Harvey Krahn, "Computer Skills and Use among High School and University Graduates," *Canadian Public Policy*, Vol.xv, No.2 (June 1989), p.183.
25. Coalition for Fair Wages and Working Conditions for Homeworkers, "Why Is Homework Increasing in the 1990s?" in *From the Double Day to the Endless Day*, Proceedings from the Conference on Homeworking (November 1992), ed. Linzi Manicom (Ottawa: Canadian Centre for Policy Alternatives, 1992), p.13.
26. Rene Morissette and Deborah Sunter, *What Is Happening to Weekly Hours Worked in Canada?* (Ottawa: Statistics Canada, Business and Labour Market Analysis Branch, 1994), p.1.
27. Margaret Philp, "Male-Female Income Gap Widens," *The Globe and Mail*, Dec. 20, 1995.
28. Cited in Goss Gilroy, "Impacts of the Information Highway on Employment

and the Workplace," a report prepared for the Information Highway Advisory Council, Ottawa, 1995, p.47.

29. Margot Gibb-Clark, "Wages Severed from Productivity," *The Globe and Mail*, Oct. 25, 1995.
30. Cordell, "New Taxes for a New Economy," p.4.
31. McLuhan, *Understanding Media*, p.51.
32. On the issue of older second-hand technology, see Glen Williams, *Not for Export: Towards a Political Economy of Canada's Arrested Industrialization* (Toronto: McClelland and Stewart, 1987), p.26.
33. Canadian Centre for Policy Alternatives, "Living with FTA/NAFTA," unpublished report, Ottawa, 1994, p.3.
34. Bruce Campbell, with Andrew Jackson, *Free Trade: Destroyer of Jobs* (Ottawa: Canadian Centre for Policy Alternatives, 1993), p.13.
35. Theresa Healy, "Selected Plant Closures and Production Relocations: January, 1989 to June, 1992, Ontario," Appendix 1, in *Canada under Free Trade*, ed. Duncan Cameron and Mel Watkins (Toronto: James Lorimer and Co., 1993), p.287; and Bruce Campbell (with Andrew Jackson), *Free Trade*, p.18.
36. Figures are from Statistics Canada, in Canadian Centre for Policy Alternatives, "Living with FTA/NAFTA," p.3.
37. Reported in Jeff Sallot, "Forging Links to Global Economies," *The Globe and Mail*, Dec. 9, 1994.
38. Bruce Campbell, "Continental Corporate Economics" in *Canada under Free Trade*, p.33.
39. Stephen Clarkson, "Constitutionalizing the Canadian-American Relationship" in *Canada under Free Trade*, p.18.
40. Student discussion panel with Peter Gzowski, "Morningside," CBC Radio, Oct. 19, 1995.
41. Grace Aiton, "The Selling of Paupers by Public Auction in Sussex Parish," Collection of the New Brunswick Historical Society, 1961, p.105.
42. John Dillon, *Economic Justice Report*, 1994, p.2.
43. Antonia Maioni, "Ideology and Process in the Politics of Social Reform," in *A New Social Vision for Canada? Perspectives on the Federal Discussion Paper on Social Security Reform*, ed. K. Banting and K. Battle (Kingston: Queen's University and Caledon Institute, 1995), pp.117-23.
44. Allan MacEachen and Jean-Robert Gauthier, co-chairs, "Canada's Foreign Policy: Principles and Priorities for the Future," Report of the Special Joint Committee of the Senate and House of Commons Reviewing Canadian Foreign Policy, Ottawa, November 1994, p.1.
45. CBC Radio, "The World at 8," CBO, Nov. 17, 1994.
46. Peter Cook, "Leaders Launch Ambitious Financial-Reform Package," *The Globe and Mail*, June 17, 1995, p.A6.

47. Tom Naylor et al., *Bleeding the Patient: The Debt/Deficit Hoax Exposed* (Ottawa: Canadian Centre for Policy Alternatives, 1994), p.6.
48. Alana Kainz, *The Ottawa Citizen*, May 28, 1994.
49. W.H. Davidow and M.S. Malone, *The Virtual Corporation* (New York: HarperCollins, 1992), p.77.
50. The quoted phrase is from R.G. Asava and R.L. Engwall, *Key Need Areas for Integrating the Agile Virtual Enterprise* (Bethlehem, Penn.: Agility Forum, 1994), p.37.
51. McLuhan, *Understanding Media*, p.56.
52. George Grant, *Technology and Empire* (Toronto: House of Anansi, 1969), pp.34, 40.
53. Geoffrey Rowan, "Reality Bites Information Technology Industry," *The Globe and Mail*, April 17, 1995; and Barbara Wade Rose, "Danger: Software at Work," *Report on Business Magazine*, March 1995.
54. Barrie McKenna, "Business Putting Bucks into Technology," *The Globe and Mail*, Sept. 24, 1994.
55. Alanna Mitchell, "Nearly Half of Workers Using Computers on the Job," *The Globe and Mail*, June 7, 1995, p.A3.

3. Hype and the Highway: Virtual Corporations and the Agile Workforce

1. Information Highway Advisory Council (IHAC), *Connection, Community, Content: The Challenge of the Information Highway* (Ottawa: Supply and Services Canada, 1995), p.x.
2. See Canada, *Telecommunications Act*, Ottawa, 1993, section 7.f: "to foster increased reliance on market forces for the provision of telecommunications services."
3. IHAC, *Connection, Community, Content*, p.xviii.
4. Ibid., pp.93-94.
5. Private communication with Industry Canada official.
6. Government of Canada, *The Internet: A Guide to Internet Use in the Federal Government* (Ottawa: Information, Communications and Security Policy Division, Financial and Information Management Branch, Treasury Board Secretariat, 1995), p.9.
7. Industry Canada, *The Canadian Information Highway: Building Canada's Information and Communications Infrastructure* (Ottawa: Supply and Services Canada, 1994), p.8.
8. Saul, *Voltaire's Bastards*, p.377. U.S. corporate debt is $2.2 trillion. Interest

payments on this absorb 32 per cent of the United States' total corporate cash flow.

9. Peter Coy, "How Do You Build an Information Highway?" *Business Week*, Sept. 16, 1991.
10. H. Dordick, H. Bradley, and B. Nanus, *The Emerging National Marketplace* (Norwood, N.J.: Ablex Publishing, 1981).
11. I disagree with those who say that the postindustrial society is an "information society." It is certainly information-intensive, but it's still industrialism. It is a new cybernetic phase of industrialism, which some have called superindustrialism, and it is emerging as the centrepiece of U.S. economic renewal.
12. Canadian Union of Public Employees (CUPE), *Computer-related Change in the Workplace*, Ottawa, 1985.
13. Pat Bird and Jo Lee, *A High-Traffic Area: Today's Automated Office* (Toronto: Times Change Women's Employment Service, 1987).
14. Patricia McDermott, *The Differential Impact of Computerization on Office Workers: A Qualitative Investigation of "Screen-based" and "Screen-assisted" VDT Users* (Toronto: Ontario Public Service Employees Union, 1987).
15. M. Cohen and M. White, *Taking Control of Our Future: Clerical Workers and the New Technology* (Vancouver: Women's Skill Development Society, 1987), pp.73, 78, 82.
16. Ibid.
17. Telephone interview with Maurice Levigne, Dec. 1-2, 1994.
18. Cohen and White, *Taking Control of Our Future.*
19. Joseph Weizenbaum, *Computer Power and Human Reason: From Judgment to Calculation* (San Francisco: W.H.Freeman and Co., 1976).
20. A. Billette and M. Cantin, *Phases in the Evolution of Computer Systems and Their Impact on the Organization of Work in General Insurance* (Québec: Université Laval, 1986).
21. B.C. Federation of Labour, *Insurance Corporation of B.C. Case Study* (Vancouver: B.C. Federation of Labour, 1985).
22. McDermott, *Differential Impact of Computerization*, p.32. The following quotes are from pp.33, 49, 35, 38 of this report.
23. Armine Yalnizyan, *The Impact of Technological Change on Hospital Workers* (Toronto: Service Employees International Union, Local 204, 1986).
24. The system used in Ottawa's largest hospital, the Civic, was developed and implemented by McDonnell Douglas. The Toronto General Hospital bought a $12 million system called Ulticare from the U.S.-based Health Data Sciences.
25. Yalnizyan, *The Impact of Technological Change on Hospital Workers.*
26. Marie-Andrée Lambert, *Etudes des impacts des changements technologiques dans*

les laboratoires cliniques des Hôpitaux du Québec. (L'Association Professionnelles des Technologistes Medicaux du Québec, 1985).

27. Kay Desborough, *The Impact of Computers on Nurses and the Care of Patients* (Toronto: National Federation of Nurses, 1985).
28. See G.J. Mulgan, *Communication and Control: Networks and the New Economies of Communication* (Cambridge, U.K.: Polity Press, 1991), p.235; and Davidow and Malone, *Virtual Corporation,* pp.91, 125.
29. R. Asava and R. Engwall, *Key Need Areas for Integrating the Agile Virtual Enterprise* (Bethlehem, Penn.: Agility Forum, 1994), p.27.
30. Comment made by Pat Armstrong during a conference on the new global economy, Carleton University, Ottawa, January 1995.
31. Armstrong et al., "Voices from the Ward," in *Take Care.* The following quotes come from this study.
32. Larry Stoffman, *Economic Context and Implications of Scanner Introduction* (Vancouver: UFCW Retail Clerks' Union, Local 1518, 1985).
33. Davidow and Malone, *Virtual Corporation,* pp.56, 58.
34. Ibid., pp.46, 48.
35. Vincent Mosco, "The Political Economy of Communication: Lessons from the Founders" in ed. Robert Babe, *Information, Communication and Economics* (Boston: Kluwer Academic Publishers, 1994), p.115.
36. Dr. Steven Goldman, *Agile Manufacturing: A New Production Paradigm for Society* (Bethlehem, Penn.: The Agility Forum, 1993), p.27.
37. Ibid., p.1.
38. Ibid., p.28.
39. Ibid., p.110.
40. Davidow and Malone, *Virtual Corporation,* p.78.
41. Mulgan, *Communication and Control,* p.27.
42. On "invisible structures," see Steven Goldman, *Agile Competition: The Next Business Paradigm* (Bethlehem, Penn.: Agility Forum, 1994), p.4.
43. Ibid., p.5.
44. Davidow and Malone, *Virtual Corporation,* p.140.
45. Asava and Engwall, *Key Need Areas,* p.41.
46. Ibid., p.17.
47. Ibid., *Key Need Areas,* p.26.
48. Ibid., pp.17, 18.
49. Betcherman et al., *Canadian Workplace in Transition,* p.41.

4. *Across the Digital Divide: Manufacturing as Global Agility*

1. "The Proven Path to Outperformance," ad for Trimark Mutual Funds, *The Ottawa Citizen* and *The Globe and Mail*, Jan. 24, 1995. For more on "Taylorism," see Chapters 1 and, especially, 5.
2. *Philadelphia Inquirer*, July 8, 1991; cited in *Agility Report*, Vol.1 (Bethlehem, Penn.: Agility Forum, 1991), p.28.
3. Ibid., p.8.
4. Goldman, *Agile Competition*, p.3; and Davidow and Malone, *Virtual Corporation*, p.154.
5. Goldman, *Agile Competition*, p.2.
6. Steven Goldman and Kenneth Preiss, eds., *21st. Century Manufacturing Enterprise Strategy*, Vol.1 (Bethlehem, Penn.: Agility Forum, 1991), p.34.
7. Gregg Jones, "Genesis of a Bomb: TI's Role Critical in Quick Development of Weapon," *Dallas Morning News*, June 30, 1991.
8. Goldman and Preiss, *21st. Century Manufacturing Enterprise Strategy*, Vol.1, pp.11, 15.
9. Heather Menzies, *Fastforward and Out of Control* (Toronto: Macmillan of Canada, 1989), p.127.
10. Bruce Roberts, "From Lean Production to Agile Manufacturing: A New Round of Quicker, Cheaper and Better," CAW Technology Project, Toronto, 1994, p.4.
11. Goldman and Preiss, *21st. Century Manufacturing Enterprise Strategy*, Vol.1, p.15.
12. R. Cooper, "Cost Classification in Unit-Based and Activity-Based Manufacturing Cost Systems," *Journal of Cost Management for the Manufacturing Industry*, Vol.4, No.3 (Fall 1990), pp.4-14.
13. "Jack Evens Says Oshawa Is on Track," *Auto Info*, Feb. 5, 1993.
14. "New Riviera/Aurora First Step in GM's Agile Manufacturing Strategy," *Ward's Automotive Reports*, Vol.68, No.49 (Dec. 6, 1993), p.1.
15. "Integrated Scheduling Project to Be Introduced at Autoplex," *Auto Info*, Jan. 21, 1993.
16. Telephone interview with Rick Dove, January 1995.
17. Interview with Bruce Roberts, Toronto, November 1994.
18. Christopher Sawyer, "Up and Running," *Automotive Industries*, February 1994, pp.124, 126.
19. Interview with Bruce Roberts, November 1994.
20. David F. Noble, *Progress without People: New Technology, Unemployment, and the Message of Resistance* (Toronto: Between the Lines, 1995), pp.78-79.
21. Cited in Noble, "Social Choices in Machine Design," in *Social Shaping of Technology*, ed. MacKenzie and Wajcman, p.109.

22. John Price, "Lean Production at Suzuki and Toyota: A Historical Perspective," *Studies in Political Economy*, No.45 (Fall 1994), p.66.

23. P.A. Lapointe, M. May, and C. St. Pierre, *Analyse sociologique des changements technologiques et l'organization du travail dans le secteur aluminium* (Fédération des syndicats du secteur de l'aluminium inc., 1987).

24. International Woodworkers of America, *The Impact of Technological Change on Productivity and Employment in the Softwood Industry of B.C.* (Vancouver: IWA, 1985).

25. "The Canadian Pulp and Paper Industry: A Focus on Human Resources," Dept. of Supply and Services, Ottawa, 1994.

26. Pierre Richard, *Etudes sur les modes d'information des changement technologiques dans l'industrie du bois au Québec* (Fraternité nationale des charpentiers-menuisiers, forestiers and travailleurs d'usines, 1985).

27. David Fairey, *Technological Change in the B.C. Fish Processing Industry* (Vancouver: United Fishermen and Allied Workers, 1986).

28. British Columbia Federation of Labour, *Grain Terminals Case Study*, Vancouver, 1987.

29. Joel Novek, *Grain Handling in the Information Age* (Winnipeg: University of Winnipeg, 1985).

30. Charlene Gannage, *The Impact of Technological Change on Toronto Ladies Clothing Workers* (Toronto: International Ladies Garment Workers' Union, 1985).

31. Steven Goldman, *Agile Competitive Behaviour: Examples from Industry* (Bethlehem, Penn.: Agility Forum, 1994), p.27.

32. John Deverell, "Computer Jeans a Fitting Idea," *The Toronto Star*, Nov. 11, 1994.

33. Coalition for Fair Wages and Working Conditions for Homeworkers, "Why Is Homework Increasing in the 1990s?" p.13.

34. Berzabeth Corona (p.5) re Mexico and Ligia Orozco re Nicaragua in *From the Double Day to the Endless Day*. "Of the six factories in the garment sector, four of them have been shut down. The women . . . are now part of the informal economy. Others are working from their homes selling whatever they can to subsist."

35. Barbara Cameron and Teresa Mak, "Chinese-Speaking Homeworkers in Toronto," summarized in *From the Double Day to the Endless Day*, p.20.

36. British Columbia Federation of Labour, *Canadian National Railways Case Study*, Vancouver, 1986.

37. Oliver Bertin, "CN Slashing More Jobs," *The Globe and Mail*, August 30, 1995.

38. Canadian Centre for Policy Alternatives, *CCPA Monitor*, July/August, 1994.

39. Jamie Swift, "Report on Northern Telecom for Canadian Auto Workers," internal publication, 1994, p.12.

40. Swift, "Report on Northern Telecom," p.11.
41. David Robertson and Jeff Wareham, *Computer Automation and Technological Change: Northern Telecom* (Toronto: CAW, 1988).
42. Ibid., p.210.
43. Karen Hadley, "Working Lean and Mean: A Gendered Experience of Restructuring in an Electronics Manufacturing Plant," PhD thesis, Graduate Department of Education, University of Toronto, 1994, endnote 16, p.96.
44. Ibid., pp.83 ,74.
45. Ibid., pp.181, 182.
46. Ibid., p.182.
47. David Robertson and Jeff Wareham, *Technological Change in the Auto Industry* (Toronto: CAW, 1988).
48. David Robertson and Jeff Wareham, *The CAW Aerospace Technology Project*, Vol.II (Toronto: CAW, 1989), p.29.
49. Ibid., p.9.
50. Ibid., pp.6, 7, 32.
51. Ibid., p.30.
52. Ibid., pp.41, 42.
53. Robertson and Wareham, *CAW Aerospace Technology Report*, McDonnell Douglas subreport, p.25.
54. Ibid., p.10.
55. Ibid., pp.14, 13.
56. Ibid., p.34.
57. Betcherman et al., *Canadian Workplace in Transition*, p.37.
58. Don Wells, interview with Jamie Swift, *The Brave New World of Work*, CBC Ideas Series, June 15, 22, 29, 1994 (transcript from CBC Radio Works, 1994), p.19.
59. Noam Chomsky, comment in press kit materials produced for *Manufacturing Consent: Noam Chomsky and the Media*, a film by Mark Achbar and Peter Wintonick, 1992; quoted from Hadley, "Working Lean and Mean," p.194.
60. Hadley, "Working Lean and Mean," p.216.
61. Donald MacKenzie and Graham Spinardi, "Tacit Knowledge, Weapons Design and the Uninvention of Nuclear Weapons," unpublished paper, University of Edinburgh, 1994, p.2.
62. Ursula Franklin, "Editorial," *Canadian Woman Studies/les cahiers de la femme*, Vol.5, No.4 (1984), p.2.
63. Joan Newman Kuyek, *The Phone Book: Working at the Bell* (Toronto: Between the Lines, 1979), p.35.
64. Ibid., p.29.

65. Franklin, *Real World of Technology*, p.107.
66. Martin, *"Hello, Central?"*, pp.98, 115.
67. MacKenzie and Spinardi, "Tacit Knowledge," p.42.
68. Hadley, "Working Lean and Mean," p.86.
69. Ibid., p.89.
70. Ibid., p.226.
71. The phrase was coined by Bruno Latour in his *Science in Action: How to Follow Scientists and Engineers through Society* (Milton Keynes, England: Open University Press, 1987).
72. Hadley, "Working Lean and Mean," p.222.
73. The term is Japanese for "suggestion." It tends to be used interchangeably with "kaisen," which is Japanese for "continuous improvement."
74. Cathy Austin, in Laura Sky (writer and director), *Working Lean: Challenging Work Restructuring*, a film produced by Skyworks for the Canadian Auto Workers, 1990.
75. Interview, December 1994, Ingersoll, Ont.
76. Hadley, "Working Lean and Mean," p.219.
77. Ibid., p.224.

5. Panopticons and Telework: The New Cybernetics of Labour

1. Nicholas Bannister, "Is This the Modern Sweatshop?" *The Gazette* (Montreal), Nov. 19, 1994, reprinted from *The Guardian*.
2. Marshall McLuhan, *Counterblast* (Toronto: McClelland and Stewart, 1969), p.142.
3. Confidential source in Ontario Premier's office, March 1995.
4. Statistics Canada, "General Social Survey and Household Facilities and Equipment Survey," Ottawa, 1995.
5. Kathleen Christensen, "Computing the Effects of Homework: Telework in the U.S.," in *From the Double Day to the Endless Day*, p.9.
6. Paul DeLottinville, *Shifting to the New Economy: Call Centres and Beyond* (Toronto: Copp Clark Longman, 1994), p.67.
7. Ibid., p.60.
8. Jerri Fowler, "Telework the New Brunswick Way," text of speech to "Telework '94 Symposium," Toronto, Nov. 15, 1994.
9. Margaret Oldfield, "The Electronic Cottage—Boon or Bane for Women?" paper presented at seminar on Gender and Economic Restructuring, University of Waterloo, May 1991, p.8.

10. Greg Weston, "Fisheries Casts off $6M on New Computer Toys," *The Ottawa Citizen*, Jan. 31, 1995.
11. Margaret Oldfield, "Heaven or Hell: Telework and Self-Employment," *Our Times*, Vol.14, No.2 (May/June 1995), p.19.
12. Bruce Roberts, CAW, personal conversation.
13. In Sky, *Working Lean.*
14. DeLottinville, *Shifting to the New Economy*, p.33.
15. Canadian Union of Public Employees, *Computer-Related Change in the Workplace*, Ottawa, 1985, p.73.
16. Larry Stoffman, "Economic Context and Implications of Scanner Introduction," UFCW, Vancouver, 1985, p.176.
17. C.A. Higgins, "Implications of Computerized Performance Monitoring and Control Systems: Perceptions of the Canadian Service Sector Worker," University of Western Ontario, London, 1987.
18. B.C. Federation of Labour, Insurance Corporation of B.C. Case Study (Vancouver: B.C. Federation of Labour, 1985). A later quote about "sloughing off" is also from this study.
19. McDermott, *Differential Impact of Computerization on Office Workers.*
20. David Robertson and Jeff Wareham, *Technological Change: Air Canada Customer Sales and Service* (Toronto: CAW, 1990), p.16.
21. See Higgins, "Implications of Computerized Performance Monitoring and Control Systems," p.40.
22. Lowe and Krahn, "Computer Skills and Use," pp.175-88.
23. Michael Valpy, *The Globe and Mail*, April 6, 1994. Considerable other evidence is collected in Maude Barlow and Heather-Jane Robertson, *Class Warfare* (Toronto: Key Porter Books, 1994), a scathing debunking of the twin notions that Canada faces a skill shortage and that the public education system is failing to deliver the skilled workers industry needs.
24. Swift, *Brave New World of Work.* See also Jamie Swift, *Wheel of Fortune: Work and Life in the Age of Falling Expectations* (Toronto: Between the Lines, 1995).
25. Mark Poster, *The Mode of Information: Poststructuralism and Social Context* (Cambridge, U.K.: Polity Press, 1990), p.58.
26. Rick Dove, *Beginning the Agile Journey* (Oakland, Cal.: Paradigm Shift International, 1993), p.11.
27. In Hadley, "Working Lean and Mean," p.186. The following quote is from the same source.
28. Mulgan, *Communication and Control*, p.86.
29. Poster, *Mode of Information*, p.90.
30. Hadley, "Working Lean and Mean," p.228.
31. Ibid.

32. Armstrong et al., in *Take Care*, p.69.
33. Telephone interview, October 1994.
34. Penny Gurstein, quoted in Oldfield, "Heaven or Hell," p.18.
35. Ibid.
36. Vivian Smith, "Repetitive Strain Injury Can Get Really Awful Really Fast," *Media*, Vol.1, No.2 (1994), p.11.
37. Sharon Saunders and Lois Weninger, "An Epidemic in the Workplace: Repetitive Strain Injuries," *Healthsharing*, Spring/Summer 1992, p.44.
38. Laura Sky (director), *The Double Edged Sword*, a film produced by Skyworks for the CAW, 1990.
39. Interview, Ingersoll, Ont., 1994.
40. Saunders and Weninger, "Epidemic in the Workplace," p.44.
41. Craig McInnes and Kevin Cox, "UPS Deal Sparks Provincial Feud," *The Globe and Mail*, Jan. 12, 1995.
42. Richard Wells, "Job Modification: Checkout Challenge," *OH&S Canada*, July/August 1993, p.64.
43. Robert Karask and Tores Theorell, *Healthy Work: Stress, Productivity and the Reconstruction of Working Life* (New York: Basic Books, 1990), pp.9, 16, 24.
44. Janine MacDonald, "Fragrance Technologies and Stress Management," final essay submitted for a Canadian Studies course, Carleton University, Ottawa, 1995.

6. A Plea for Time—A Plea for Our Times

1. An abbreviated version of this story first appeared in Menzies, *Fastforward and Out of Control*.
2. Harold Innis, *The Bias of Communication* (Toronto: University of Toronto Press, 1984), pp.61-91.
3. For the spoken word as site, see Andrew Wernick, "No Future: Innis, Time-Sense and Postmodernity," in *Harold Innis in the New Century: Reflections and Refractions*, ed. C.R. Acland and W.J. Buxton (Montreal and Kingston: McGill-Queen's University Press, forthcoming).
4. I wish to thank Bruce Roberts of the Canadian Auto Workers for clarifying for me that unions are an oral culture.
5. Sky, *Working Lean*; and Armstrong et al., "Voices from the Ward."
6. CAW, *Workplace Issues: Work Reorganization: Responding to Lean Production*, Toronto, 1994, p.16.
7. Vandana Shiva, *Monocultures of the Mind* (London: Zed Books, 1993), p.62.

7. On Our Own Terms

1. Carey, "Canadian Communications Theory," p.32.
2. Ibid., p.33.
3. Ibid., p.32.
4. Re pay-per-use, a 1995 Bell pamphlet advertises "three-way calling" as "a pay-per-use service."
5. James Carey, *Communication as Culture: Essays on Media and Society* (London: Unwin Hyman, 1989), p.18.
6. This is a variation on something Ursula Franklin told me in conversation: "People don't want more information; they want relationships."
7. Part of this is drawn from a keynote speech I gave to the Atlantic Regional Libraries Association in May 1995 (forthcoming). See also my essay, "Information Gathering and Confidentiality: Data Bases, Monopolies of Knowledge and the Right to Be Informed on Your Own Terms," in *Human Rights in the Twenty-First Century: A Global Challenge*, ed. K. Mahoney and P. Mahoney (Boston: Kluwe Academic Publishers, 1993).
8. IHAC, *Connection, Community, Content.*
9. Stan Skrzeszewski and Maureen Cubberley, *Future-Knowledge: A Public Policy Framework for the Information Highway* (Toronto: Ontario Library Association's Coalition for Public Information, 1995), p.4.
10. Andrew Reddick, *The Information Superhighway: Will Some Canadians Be Left on the Side of the Road?* (Ottawa: Public Interest Advocacy Centre, 1995), p.5.
11. Andrew Reddick, *Sharing the Road: Convergence and the Information Highway* (Ottawa: Public Interest Advocacy Centre, 1995), p.14.
12. Ibid.
13. Freedman, "Report of the Industrial Inquiry Commission," p.95.
14. Ian Austen, "Trains Start Running Today," *The Ottawa Citizen*, March 27, 1995.
15. Vincent Mosco. *Doing it Right with Computer-Communication: A Case Study of the United Services Auto Association* (Boston: CIPR, Harvard University, 1994).
16. Joey Hartman, "Worker Initiatives for Alternative Technological Applications," Vancouver Municipal and Regional Employees' Union, 1987.
17. Gay Abbate, "Ottawa Urged to Cap Workweek, Paid Overtime," *The Globe and Mail*, Dec. 17, 1994; Virginia Galt, "Longer Work Week Widens Wage Gap Between Rich, Poor," *The Globe and Mail*, July 13, 1994; Margot Gibb-Clark, "Jobs Keeping Couples Apart," *The Globe and Mail*, June 2, 1995.
18. Paul Willich, "A Workaholic Economy," *Scientific American*, August 1994, p.89.

19. Women's Bureau, Human Resources Development Canada, *Women and Economic Restructuring: Report of the Committee on Women and Economic Restructuring* (Ottawa: Canadian Labour Market and Productivity Centre, 1994), p.13.
20. Morissette and Sunter, *What Is Happening to Weekly Hours Worked in Canada?*, pp.14, 15.
21. Interview with Swift, "Brave New World of Work," p.28; see also Swift, *Wheel of Fortune*, pp.228-29.
22. Steven Langdon, *The Emergence of the Canadian Working Class Movement, 1845-1875* (Toronto: New Hogtown Press, 1975), p.13.
23. Ursula Franklin has called for a census on machines since chairing a Science Council study on a conserver society in the 1970s.
24. James Tobin, "Speculators' Tax," in *New Economy* (Fort Worth: The Dryden Press, 1994).
25. Adrian Randall, "The West of England Wool Workers—1790-1809," *Technology and Culture*, Vol.27, No.1 (1986).
26. Peter Mayer, ed., *The Pacifist Conscience* (New York: Holt, Rinehart and Winston, 1966), pp.394, 395, 398.
27. The Alliance for a Connected Canada includes the Telecommunications Workers' Union, the Communications, Energy, and Paperworkers Union, the Council of Canadians, the Public Interest Advocacy Centre, the Coalition for Public Information, and Telecommunities Canada. The affordable telephone coalition represents over sixty consumer, seniors' community, women's, students', and antipoverty groups.
28. See Penny Kome, *The Taking of Twenty-Eight: Women Challenge the Constitution* (Toronto: The Women's Press, 1983).
29. John Dewey quoted in James Carey, *Communication as Culture*, p.14.

Index

ABC (T.V.) 57
Action Canada Network 141
ACTRA 158
actual working time (AWT) 114
advertising 18, 25, 44, 123, 124
agile factories 22, 164
agile manufacturing 80, 85, 91
agile workers 71, 102
agile workforce 80, 81
See also credentialism *and* virtual corporations
Agility Forum 74, 76, 85, 124
Air Canada 120, 121
Alliance for a Connected Canada 162
American Airlines 72
Angus Reid poll 10
anti-trust laws 75
AT&T 55, 77, 93
automation 21, 29, 60, 89, 98, 99.
See also computerization *and* restructuring

Bank of Montreal 135
Baudrillard, Jean 123
B.C. Packers 89
Bélanger, Marc 141
Bell, Alexander Graham 147
Bell Canada xiii, 5, 6, 10, 55, 72, 77, 93, 105, 126, 135, 137, 161
Bellemare, Diane 26
Benetton 71, 81, 91
Bentham, Jeremy 124
bias of communication 22, 138, 146
Big Brother 4, 18, 19, 118
Bissonette, Sophie 47, 126
bit tax 11
See also Cordell, Arthur
blue-collar work 80.
See also employment trends *and* restructuring
Boeing 100
Brampton 93, 95, 102
branch-plant industrialization 37, 38
Brantford 105
British Columbia 52, 61, 64, 70, 89, 129
Budapest 105
Burton, Donald 3
Business Week 57, 70

call centres 4, 6, 10, 22, 37, 51, 114, 116, 118, 129, 142, 144, 164.
See also information highway *and* telework
Camco 115
CAMI 103, 125, 129, 154
Campbell, Bruce 38
Canada 74.
See also Federal Government
Canadian Auto Workers 93, 97, 99, 103; and competitiveness 140
Canadian Centre for Policy Alternatives 15, 38, 141
Canadian Charter of Rights and Freedoms 163
Canadian Congress for Learning Opportunities for Women 156
Canadian filmmakers 149
Canadian Labour Congress (CLC) 156

Canadian National Railways 23, 92, 161
Canadian Pacific 93, 161
Canadian Radio-television & Telecommunications Commission (CRTC) 135, 136
CANARIE 52
Carey, James 15, 147
Cargill 90
C.D. Howe Institute 15
childcare 92
child labour 92, 160
China 91
Chomsky, Noam 104
Chrétien, Jean 41, 52
Chrysler Canada 33
citizens 12, 41, 134, 138;
census of computers 159
civil society 134, 164
Clarkson, Stephen 39
class,
divisions 18, 32, 110, 124;
middle 5, 32, 36, 75, 78
colonization 40, 53, 56, 77, 78, 100
communication networks 73, 164;
500-channel universe 51, 52, 54, 55, 57, 109;
corporate 7, 21, 35, 58, 74, 76;
models/theories of 53, 55, 136, 137, 145, 150, 153, 162;
network services 58;
structures of 59, 144, 152.
See also freenets *and* information highway
Communication policy (mixed model) 155, 163
See also Information Highway Advisory Council
Communication Workers Union 135
communications days 104.
See also training
community 59, 111, 119, 134, 141, 153;
service groups 161.
See communication networks *and* workfare
community-information services 28, 136.
See also freenets
competitiveness/global competition 16, 23, 25, 54, 140, 142;
and information highway 56
Computer and Business Equipment Manufacturers' Association 121
computer control 9, 11, 12, 21, 36, 59, 87, 144;
access to 89, 96, 100, 103;
as experience 134, 147;
centralization/concentration of 102, 151;
computer monitoring and performance measurement 9, 12, 31, 64, 70, 91, 108, 117-122, 127;
gender difference 65, 118;
struggle over 64, 88, 90, 108.
See also cybernetics of labour
computer literacy/skills 33, 39, 47, 95;
and underemployment 33, 122
computer simplification 9, 51, 88, 95, 122.
See also McJobs
computer simulation 71
computerization xiii, 21, 36, 43, 44, 63, 112;
closure 36, 44, 47, 58, 66, 113, 117;
cultural impacts of 36, 44, 70, 88, 147, 159
integration phase 21, 32, 38, 58, 59, 61, 83, 89;
networking phase 21, 29, 30, 37, 58, 63, 86, 144.
See also employment trends, labour force, *and* restructuring
concurrent production 74.
See also restructuring
Conference Board of Canada 8, 35
conserver society 158
See also knowledge society
Consumers Gas 115
contingent labour force 59, 75, 76, 77, 84, 85, 110, 115;
status 134.
See also employment trends *and* labour force

contracting out 40, 60, 89, 98, 99.
See also downsizing *and* privatization
convergence 55, 57
cookie-cutter people and production 86
Cordell, Arthur 11, 35
Corel 42
corporate cultural training 12.
See also training
Corporate Information Technology Institute 123
corporatism/ist 12, 14, 41, 134, 139, 141, 162
Council of Canadians 141
credentialism 32, 34, 97
Cronkite, Walter 102
Cross-Country Check-up 148
cross-functional/enterprise teams 75, 81, 85
cross-training 97
cultural industries 55
culture 55, 147, 154;
of workplace 100, 112, 126, 127, 140.
See also ecology
CUPE 118, 141
customization 82;
customized mass production 71
cybernetic(s) xiv, 115;
apartheid 128;
of labour 32, 36, 77, 110, 144, 151;
systems 35, 90;
workplace 155
cyberspace 11, 57, 59, 71, 79;
political economy of 11, 30

Daily Bread Food Bank 5, 10, 122
Dassinger, Janet 18
de Havilland 99, 100
de-institutionalization,
of government 164;
of workers 31, 37, 59, 86, 156.
See also contracting out, privatization, *and* virtual corporations
de-unionization 31, 37, 60, 101, 156
declining middle 9, 30, 32, 36, 97.
See also class, employment trends, *and* middle class
deficit/debt 5, 25, 143;
cutting 7, 16, 25, 37, 40, 155;
origins 25, 26.
See also social programs
deficit spending 26
Depression 4
deregulation 25, 38, 90, 142
deskilling 18, 34, 35, 61, 88, 89, 94, 97, 98.
See also cybernetics of labour, employment trends, *and* labour force
deskilling/reskilling debate 94, 100
Dieppe (N.B.) 116
digital distribution (of work, goods and services)7, 10, 22, 37, 60, 71, 77, 78
digital divide 10, 87, 110.
See also silicon curtain
digitization,
of economy 20;
of work 101.
See also automation *and* restructuring
discourse xiii, xiv, 4, 15, 18;
feminist 139;
new critical discourse 17, 45, 47,134, 137, 140, 151, 154, 163, 164;
official on restructuring 45, 46, 133, 140, 153;
on nuclear arms 138;
and official reality 5, 14, 15, 16, 141, 142.
See also restructuring
Dolci, Danilo 161
Dove, Rick 124
downsizing 8, 17, 78, 93, 134, 155, 159
language of 138.
See also employment trends *and* welfare state
Drucker, Peter 80
Dupont 71
Dvorak design of keyboard 27

Eaton's 81
ecology (social) 12;
of work 112, 120, 128
Economic Council of Canada 25

economic renewal 52;
corporate 53
economies of scale, scope, speed 40.
See also scale
economy,
global economy 13, 37;
industrial 9, 37, 58;
national 37;
post-industrial xv, 13, 58, 74.
See also new economy *and* restructuring
education,
and jobs/income 33;
declining enrolments 39;
restructuring in 154;
spending 26, 39;
tuition 26, 39.
See also Schoolnet *and* training
Eid, Ed 42
El Salvador 91
Ellul, Jacques 23, 133
employment centres 8, 53
employment trends 6, 9;
good jobs/bad jobs 9, 31, 36;
job loss 30, 89, 90-3, 98, 101, 143;
job gains 6, 95;
jobless economic growth 31, 78;
part-time/temporary 5, 34, 59, 60, 62, 66, 67, 70, 95;
polarization of jobs, hours and wages 9, 10, 30, 32, 33, 61, 88, 156, 157;
self-employment 158;
skill requirements 31, 95;
two-tiering 18, 30, 62, 95.
See also contingent labour force, deskilling, *and* jobless economic growth
empowerment 107, 124, 154.
See also TQM
environment 51;
cultural/social 128, 134, 155;
computerized/digital 58, 59
ergonomic 92
excellence 124.
See also team excellence *and* TQM
Excello Colonial 98
expert(s) xiv, 5, 16, 17, 24, 43, 45, 133, 134, 137, 153.
See also corporatism *and* discourse
expert systems/software 31, 61, 65, 78;
as calculation 63

Factory America Network 73, 85
Federal Express 114
Federal Government,
and Internet 55;
and telework 116;
back-to-work legislation 152;
conference on telework and the information highway 55, 114;
layoffs 6, 8, 159, 161;
spending on information technology 8, 53;
spending on information highway 52;
structural adjustment 133;
Telecommunications Act 54.
See also CANARIE *and* Information Highway
feudalism xiv
flexible manufacturing 85, 102.
See also agile manufacturing
Food Marketing Institute 70
Ford 103, 104
Fordist social contract 4, 53;
work model 88
Foucault, Michel 117
Fowler, Jerri 114, 116
Franklin, Ursula 20, 29, 46, 96, 105, 106
Fraser Institute 15
free trade 25, 26, 37, 38, 39, 40, 78, 79, 90, 97, 142;
Freedman, Samuel 23, 143, 152
freenets 28, 55, 59, 136, 148, 155, 156, 162
full employment 162, 163

G7 Summit 41
gender,
exclusions 124.
See also computer control, men, sexism, *and* women
General Electric 57, 115, 149;

General Electric Information Services Co. 57, 71
General Motors (GM) xvi, 4, 53, 85
geopositional satellites 115, 118. *See also* surveillance
Germany 157
global economy 16, 56, 76, 137;
governance 41, 42, 74,
participation in 138.
See also restructuring *and* virtual corporations
global village 19, 134.
See also McLuhan
globalization 5, 41, 133;
corporate agenda on 140.
See also restructuring *and* virtual corporations/enterprises
The Globe and Mail xvi, 4, 14, 15, 122
Goldman, Steven 82
good jobs/bad jobs. *See* employment trends
Gore, Al 56, 57
government cutbacks 56, 79.
See also social programs
Grand National Apparel 51
Grant, George 12, 13, 20, 23, 45
Green Revolution 140
Greenpeace 156

Hadley, Karen 96, 97, 107
Harley Davidson 73
head tax 160;
on computers 159.
See also Tobin tax *and* bit tax
health 46, 62, 121;
as public resource 154.
See also stress
health care,
and restructuring 44, 56, 65-69, 118;
spending 26;
cutbacks in 47;
privatization 35, 69;
and safety 103, 121;
and women 92;
workers 6, 118, 127.
heart disease 130
Heart Fund 137
Hegel 23
home-based work 91, 92, 116.
See also women *and* telework
hospitals 56, 154
hours of work 6, 78, 156-158;
coalition on 156;
Nine-Hours League 158;
overtime and tax/benefits 157;
overtime/longer hours 56;
polarization in 33, 34;
shorter workweek 157, 158
human interface 144
Human Resources Development,
sectoral councils on 18;
Standing Commons Committee of 26
Hunnicutt, Benjamin 158
Huxley, Aldous 19

IBM 55, 62, 77
Inco 72, 83, 89
Industrial Revolution 7;
economy 58
industrialization of services 62, 67, 69
informating 153.
See also automation *and* computerization
information highway xiv, 6, 7, 12, 20, 21, 27, 29, 30, 40, 43, 51-9, 101, 103, 109, 142, 143, 144, 148, 155;
access to 54, 55, 150;
and virtual corporations/enterprises 22, 43, 86;
as axis/centre of new economy 53, 56, 58, 59, 69, 86;
as colonizing agent 53;
as distribution medium 21, 37, 144;
constitution/social charter for 162, 163;
corporate players 52, 55, 72;
discourse/debate on 22, 54;
federal government 52, 55;
history of 56, 57;
infrastructures of 20, 21, 27, 52, 54, 58, 74, 109, 163;

regulation/governance of 145, 155, 160, 163.
See also digital distribution
Information Highway Advisory Council 54, 150, 162;
report of 22, 149, 150;
minority report on 54
Ingersoll 103, 125, 129, 154
Innis, Harold 22, 56, 59, 138, 140, 146
Intel 43
intensification of work 60, 97;
of home life 73.
See also stress *and* technological dynamo
international financial markets 13
International Monetary Fund (IMF) 19, 41, 42, 47, 74
Internet 7, 10, 12, 29, 30, 47, 55, 59, 109, 142, 148;
access to 55;
business convergence with 55, 58;
entrepreneurs on 10, 144, 149

Japanese 57, 84, 102
J.C. Penney 71, 81, 115
job creation 143
job security 18.
See also employment trends
jobless growth 31, 78, 82, 88, 89, 94, 143
Johnston, David 22
joint-stock cooperatives 162
just-in-time management 70, 84

kaisen 107
Kansas City 104
Kapor, Mitchell 20
Kellogg, W.K. 157
Kenworth Truck 98
key words 16, 17, 124
Kiwanis Club 40
knowledge society xv, 12, 109, 144
Korea 102

labour force/market 9, 76;
computer skills 33, 39;
continental/global division of 10, 25, 30, 108;
education, wage,job 33;
hours/wages polarization 33, 34, 78, 92;
older workers 34, 97;
standards/benefits 5, 79, 89, 92, 158.
See also employment trends, health, *and* restructuring
labour-management relations 24.
See also computer control *and* cybernetics of labour
Laurence, Margaret 133, 138
law offices 62
layoffs 6, 30;
survivor syndrome 6, 130.
See also restructuring
leisure society 144, 158.
See also hours of work *and* knowledge society
level playing field 79.
See also free trade *and* restructuring
Levi Strauss 91
libraries 148, 154;
and information highway 78, 149;
restructuring in 60
Livingstone, David 33
local/regional economies 30, 37, 56
Lockheed 82
London (Ont.) 95, 116
Luddites 24, 160
Lumpkin, Ramona 26
Lyotard, Jean-François 21

Maclean's 14
Maioni, Antonia 41
managerial control/prerogative 23, 24, 143
maquiladores 38
Masse, Marcel 13, 109, 147
McDermott, Patricia 120
McDonald's 9
McDonnell Douglas 100
McGill University 22
McJobs 9, 19, 36, 39, 43, 62, 91, 113, 127, 144, 151.

See also cybernetics of labour *and* employment trends
McKenna, Frank 6, 15, 114
McLaren, Roy 38
McLuhan, Marshall xiv, 3, 9, 12, 13, 14, 18, 21, 23, 44, 141, 146;
global village 19, 34;
the medium is the message 51, 52, 146;
turning turtle 110, 126
McQuaig, Linda 26
Meals on Wheels 162
MEC Air Technologies 130
media xiv, 140.
See also discourse
men 35;
hours of work 33, 35;
masculine mystique 157.
See also employment trends *and* restructuring
metrics (computer) 68, 75, 111, 147.
See also computerization *and* servo-mechanism
Metro Toronto 128
Mexico 38, 74, 92, 108, 143
microprocessor, -chip 21, 29, 30;
capacity of 43, 57, 58
Microsoft 10, 28, 43, 55, 77, 149;
Windows '95 29
Middle Ages 73
middle class 5, 75, 78;
income 32.
See also class *and* declining middle
Midland 112;
telephone operators/exchange 134-7, 140, 141, 152
Millikans Industries 81
Millington, Karen 121
miniature replica 89.
See also branch plants
MIT 84
monoculture 13, 22, 87, 140, 141
monopoly,
capitalism 74;
corporate/structural 11, 73;
of knowledge/thought 15, 112.
See also scale *and* virtual corporation
moonlighting 6.
See also employment trends *and* labour force
Moore, Gordon 43;
law 43, 86,87, 88
Moses, Robert 27
Motorola 84
Mulroney government 24.
See also free trade *and* restructuring
multi-media 7.
See also information highway
multiskilling/multitasking 31, 92, 94, 98, 99, 101
Murphy's Law 46, 88

nation state 74
National Action Committee on the Status of Women (NAC) 141, 156
National Council on Welfare 15
national economy 37;
scale-up to global 56.
See also economy
National Federation of Nurses 67
National Film Board 148
National Policy 38
National Secretary's Day 46, 105
National Trustco 8
Naylor, Tom 42
NBC 58, 77
neo-conservativism 41
neo-liberalism 41
network services 58.
See also communication networks
New Brunswick 4, 6, 40, 52, 118;
Information Highway Secretariat 62, 114;
Telephone Co. 114
New York 27, 91, 160
New Democratic Party (NDP) 136
new economy 5, 11, 13, 28, 47, 51, 73, 86, 142, 143;
polarized 110;
social charter for 163.
See also restructuring *and* virtual coporations

Newspeak 4, 17, 18, 26
Nicaragua 92
Noble, David 87
non-standard jobs 34.
See employment trends
Northern Telecom 77, 93-96, 102, 116
numeric control (NC) 86, 87, 97, 98, 99
nurses 7.
See also women

Oakville 103, 104
Oldfield, Margaret 128
1-800 numbers 10, 30, 81, 115
Ontario 38, 40, 52, 66, 93, 103, 112;
and call centres 115, 126;
and de-industrialization 25, 37, 38
Ontario Conservatives 5
Ontario Institute for Studies in Education (OISE) 33
Ontario Library Association 150
Oppenheimer, Robert 23
oral communication 139, 140;
oral culture 137
Orlikow, David 136
Orwell, George 4, 17, 18, 133
Orwellian 107;
perception management 14
Oshawa 3
Ottawa 6, 123
outsourcing 17.
See also privatization
over-time 88, 101, 157;
and tax 157.
See also hours of work

panopticons 118, 125;
and cultural training 118;
Bentham on 125;
socializing effects of 126.
See also surveillance
paradigm shift 11, 43
Paris 105
Parliament 136
Parrot, Jean-Claude 54
part-time employment.
See employment trends *and* labour force
pauper auctions 40
pay equity 62.
See also women
payroll taxes 39, 79, 114
People for Affordable Telephone Service 162
Persian Gulf War 82
Phillips Electronics 55
Pizza Pizza 127, 128
plant closures 25, 30, 89, 90
politics of identity 138
post-Fordist work model 77, 101, 110.
See also telework *and* Toyotism
post-industrial society/economy xv, 13, 55, 74;
corporations 22;
renewal 82
"Post-It Note" people 10, 47, 59
post-modernists 123
post-Taylorism 103, 125.
See also Taylorism
poverty,
increase of 5;
youth 5, 34
poverty police 40
power (in society),
distribution/balance of 11, 35, 110, 143, 144, 150, 151, 152, 158, 159;
loss of (including loss of control) 88, 98, 99, 100, 107;
people- 159
Prince Rupert 90
privacy 31.
See also cybernetics of labour *and* panopticons
privatization,
from within 40, 78;
from paid to unpaid work 10, 31, 73, 89;
from public to corporate sector 7, 40, 69.
See also colonization *and* prosumers
productivity 16, 24, 30, 143
profit sharing 75

prosumers 31, 73;
definition of 81.
See also privatization *and* teleconsumers
Public Interest Advocacy Centre 150, 156
public readings 148.
See also communication networks *and* culture
public-service employment 25, 78.
See also welfare state
Purolator 114, 115

quality circles 84, 126, 154.
See also TQM
Quebec 63, 89, 93
Queen's University 103, 111
quick response (QR) 71, 77, 81, 84, 102, 115

racism 92;
and training 92, 104
railways 23
rationality/reason 13, 15
recession 5, 32, 40, 134
Reddick, Andrew 151
redundancies 61, 108, 133
re-engineering 111.
See also restructuring
Regina 3
regional/local economies.
See local/regional economies
reindustrialization 74.
See also restructuring
repetitive strain injury (RSI) 6, 108, 130;
and autonomy 129;
extent of 129;
and workers' compensation 129.
See also stress
reserve army of unemployed 31
reskilling 94, 98
restructuring xiv, xvi, 7, 11, 14, 27-32, 35, 38, 44, 82, 138;
alternative agenda on 140;
and shift to part-time 60, 61;
and taxes 159, 160;
debate about xv, 16, 17, 23, 24, 25, 45, 151, 159;
debt link 109;
in government 60;
in manufacturing 74, 83-102;
in services 59-72;
negotiation/struggle xv, 145, 153;
social context of 134;
social/cultural impacts of 30-2, 36, 59-72, 89-102, 141;
stages in 76;
welfare state 35, 41.
See also computerization
retraining 137.
See also retraining
reverse strike 161
Roberts, Bruce 103
Robertson, David 103, 118
Robillard, Lucienne 152
Rotary Club 40
Royal Bank 5, 116

Sabreline 72
Salutin, Rick 16
Saul, John Ralston 12, 15
scale 85;
economies of 40, 43, 77;
global 79;
monopoly 22, 73, 77;
-up of economies 56, 73
scanners 30, 70, 81
Schoolnet 28
Science Council 25
self-employment 151.
See also Internet
Senate-House Committee on foreign policy 41
servo-mechanism 36;
people as 13, 36, 44, 113
See also technology
sexism 32, 92;
and training 97
shiftwork 70, 94, 127
Shiva, Vandana 13, 140
Sholes, Christopher 27
silicon ceiling 34.

See also call centres *and* telework
silicon curtain 10, 31, 36, 95
Sky, Laura 47, 139
Smith, Adam 35
Smith, Dorothy 17
social contract 133, 150;
Fordist 4, 53
social economy 11, 134, 150
social justice xiv, 24, 32, 137, 142, 151
social programs 35, 163;
cuts to 7, 35, 37, 40, 134, 143
and restructuring 56
See also privatization
social safety net 33
social standards 7;
levelling of 39, 41
Solinet 141, 156, 159
South America 143
Southeast Asia 143.
See also Japan
spending cuts. *See* deficits *and* social programs
Ste. Agathe xiii, 112
Stentor 77, 117, 149
stereolithography 86, 87
stress 6, 42, 62, 65, 92, 99, 108, 120, 121;
lack of control 65, 130;
layoff-survivor syndrome 130;
management of 130;
workers' compensation 129.
Sudbury 72, 89, 105
super-panopticon 117
surveillance 12, 37, 40, 41.
See also panopticons
Sweden 90
Swift, Jamie 123

tacit knowledge 46, 47, 61, 105-108, 113, 137;
and nuclear disarmament 106;
appropriation of 102, 106;
power of 46, 107, 138, 140, 153
tax base 25, 26, 163;
and deficit 26;
and overtime 157;
burden 35
Taylor, Frederick 12, 124
Taylorism 12, 81, 101, 124;
and stress 130
TC^2, 81, 91
Team Canada 41
team excellence 106, 107.
See also TQM
technocratic 134.
See also corporatism
technological change 25, 34.
See also restructuring
technological dependency 43
technological determinism 22, 27, 45
technological dynamo 12, 42, 44, 75, 108
technological enfranchisement/disenfranchisement 11.
See also computer control *and* power
technological practice 112, 137;
women's 111.
See women
technology,
as extension of ourselves 23, 44, 149, 151;
as system 23, 27, 28, 149;
as tool 11, 26, 56, 59, 61;
holistic/growth model 29, 43, 66, 94, 100, 137;
impact on people of 24;
infrastructures of 23, 27, 29, 59;
negotiation of 24, 153;
people as extensions of 10, 36, 39, 44;
people's impact on 46;
prescriptive/production model 29, 43, 55, 83, 137;
social construction of 22, 27-9;
See also computerization, employment trends, restructuring, *and* servo-mechanism
Telecommunications Act (1993) 54.
See also Federal Government *and* information highway
telecommuting 76, 110, 115, 117
teleconsumers 143, 151
See also prosumers
telemarketing 114, 115

telephone,
history of 28;
operators xiii, 28, 105, 112, 114, 126, 134-7, 140, 141;
rates 10, 52, 147
telescreens 4, 5, 18
teleshopping 51, 71, 119
telework 76, 110;
conference on 55, 114;
extent of 47, 77, 115, 116;
labour standards 158;
personal experience of 127, 128, 134;
women and 116.
See also call centres *and* women
teleworkers 22, 37, 51, 77, 151;
and RSI 129
Texas Instruments 82
Third World 42, 98;
wages 37;
debt 42
time 13, 42, 43, 138, 148, 162;
as experience 139.
See also technological dynamo
Time Warner 77, 149
Tobin tax 160
Toffler, Alvin 31, 37, 81
tools,
of communication 11, 21, 142;
of production 11
Toronto 5, 61, 91, 105, 122, 127
Toronto Star 3
total quality management (TQM) 12, 46, 84, 102;
and computer monitoring 118, 122;
origins of 68
Tower of Babel 159.
See also computerization *and* culture
Toyotism 88, 101.
See also post-Fordist work model
training,
as industry 122, 123;
corporate cultural training 36, 102, 103, 124;
distribution of 32;
for compliance 12, 122-124;
inadequacies of 100-02;
skills training 103;
struggle over 32, 88, 102.
See also credentialism
transmission 145, 147.
See also communication networks
transnational corporations 78.
See also monopoly *and* scale
Trojan Horse 57
typewriter 27

underemployment 5, 9, 31, 33.
See also employment trends
unemployment 9, 19, 47;
rates 4, 5, 6, 30;
insurance and changes to 26, 36
unions 40, 60, 76, 127, 142, 155, 158;
and strikes 90, 107, 161;
and technological change 135, 136, 139;
business unionism 155, 156;
virtual 156, 158, 164.
See also deunionization
United Food and Commercial Workers Union 18
United Parcel Service (UPS) 4, 114, 115, 129
United Services Auto Association 153
United States 23, 63, 69, 70, 81, 83, 90, 91, 129, 143;
assets 93;
branch plants 38;
Defense Dept. 55, 74, 82;
economic liberalism 41;
free trade 37;
High-Performance Computing Act 57;
National Information Infrastructure Program 52;
Occupational Health and Safety Administration 129
Université Laval 63
University of British Columbia 128
University of Western Ontario 121
University of Windsor 26

Valpy, Michael 122
Van Helvoort, Carol 127, 128

Vancouver 152;
Municipal and Regional Employees Union 153;
municipal office workers 152, 155
Vietnam 41
virtual classroom 21
virtual clothing 91
virtual corporation/enterprise 11, 21, 69, 73, 85
virtual economy 22, 46.
See also new economy *and* post-industrial economy
virtual goods and services 71, 81
virtual workplace/site 10, 20, 37,77

Walmart 30, 71, 81, 148
war on drugs 83
war-readiness 82
Weizenbaum, Joseph 63
welfare 4, 161;
state 41
Wells, Don 104
Westinghouse 57
WHEFTA 39
White, Lillian 135
Williams, Raymond 16
Winner, Langdon 51
Witty, Grace 105
women xiii;
and constitutional conference 163;
and seniority 96;
and Studio D 148;
and tacit knowledge 105, 106;
and TQM 104, 107, 127;
and wage gap 31;
clerical workers 61, 63, 95, 122, 152, 153;
computer monitoring/control of 61, 118;
double day 97;
employment 33, 34, 61-72, 93, 96;
ghettoization 33, 34, 64, 90, 93;
home-based workers 34, 91, 92, 116;
hours of work 33, 92, 157;
immigrant women 90;
of Midland 135-36;
non-traditional work 97;
older women 65, 118;
redomestication of 35;
technological practice of 28, 105, 106, 111, 137;
telework 116
women's movement/groups 45, 138, 142, 158
work reorganization 68.
See also restructuring
workers compensation 129.
See also health
workfare 39, 40, 47, 133, 162
Working Lean 139
World Bank 109
World Wide Web 28, 144;
websites 30, 55
Writers' Union of Canada 158

York University 139
youth,
poverty of 5, 34;
unemployment/underemployment 6, 33